Heat Pumps

- Heating and Cooling -

A Beginner's Handbook

Technology, Planning to Commissioning, with Step-By-Step Guide and Checklist

Ziegler

PUBLISHED.
JUST.4U

it's your choice!

ISBN Paperback: 978-3-910707-03-0
ISBN Hardcover: 978-3-910707-04-7
ISBN E-Book: 978-3-910707-05-4

TABLE OF CONTENS

INTRODUCION

If you're looking to heat your home cost-effectively and independently while also benefiting the environment, a heat pump system might be the perfect solution for you. The heating season typically spans from early October to the end of April, covering a significant portion of the year. Additionally, there's the consistent need for hot water, irrespective of the building standard or seasons. Both heating and hot water requirements are catered for by the heating system.

So, why a heat pump? Let's summarize the primary reasons:

Resource Independence: Using a heat pump ensures independence from fossil fuels. This means that regardless of global or geopolitical shifts, such as price hikes or resource shortages, you can heat your home autonomously.

Environmental Impact: Utilizing a heat pump significantly reduces your personal CO^2 emissions, contributing to the global fight against climate change.

Cost Savings: With proper planning (detailed in Chapters 7 to 11), homeowners can realize substantial annual energy cost savings.

Future Flexibility: Heat pumps provide flexibility for integrating other alternative energy generation methods in the future, like photovoltaics or solar panels (see Chapter 10), further enhancing energy conservation.

Property Value Boost: Implementing a sustainable technology like a heat pump can significantly increase the future value of your property and help in meeting potential future environmental energy regulations.

Space and Maintenance Savings: As heat pumps typically operate on electricity, there's no need for storage space for fuels like oil. Moreover, there are no onsite emissions, eliminating the need for chimneys and associated maintenance.

With this, you're now acquainted with the pivotal reasons that have made heat pump technology increasingly essential, leading to their escalating adoption in households in recent years.

Given the rising adoption of heat pumps, this book will delve into energy-efficient heating using heat pump systems. **This book is designed for beginners. Its objective is to provide foundational knowledge on heat pump systems and equip you with the tools for planning and operating a heat pump system, thereby preventing common mistakes that can be costly.**

This is both a technical and a working guide and can be read in two different ways:

The knowledge conveyed in this book follows a chronological order, offering a comprehensive understanding and guidance for planning and implementation. You can read the book from start to finish. If you're new to heat pumps, this approach is recommended.

Alternatively, you can focus only on parts of the book that pique your interest, or select chapters based on the table of contents, without having to read the book sequentially.

The book is structured into **three main parts** and an **appendix**:

PART I: Delivers a comprehensive understanding of the basic principles of heat pump systems. Beginning with an explanation of energy fundamentals and terminologies, it ranges from the description of an efficiency house to various types of energy and finally to the definition of the (Heating) Seasonal Performance Factor crucial for heat pumps. It goes on to introduce the structure and functionality of heat pumps, highlighting different application areas and the variety of heat sources that can be utilized. This includes a thorough discussion on various extraction technologies tailored for specific heat sources. A pivotal segment distinguishes between operational modes, enabling one to attain an individual optimum during planning. Lastly, it describes the diverse types of hot water storage systems available.

PART II: Transitions to the practical application of the foundational knowledge acquired in Part I. The most extensive chapter in this part provides a step-by-step planning process for your heat pump system, utilizing a graphical planning scheme, guiding you right through to the final decision and placement of the order. Every planning step is broken down into detailed sub-chapters, with relevant calculations, supported by concise examples. Upon finalizing your heat pump choice and completing the order process, the subsequent chapters discuss installation and commissioning, further segmented into specific chapters. The concluding chapter of Part II addresses measures to ensure the long-term efficient and smooth operation of your heat pump system, focusing on inspection, maintenance, and upkeep.

PART III: Delves deeper, extending the foundational knowledge from Part I. It offers insights into refrigerants vital for the operation of heat pumps, the integration of alternative energy systems, guidelines for troubleshooting and rectifying faults, and key considerations for refining the performance of the heat pump system.

The **Appendix** encompasses indispensable tools for planning, such as the planning schema itself, a planning checklist, and the heat pump matrix, assisting in the selection of the appropriate heat pump.

Throughout the book, cross-references between chapters ensure a thorough understanding without any gaps. To cater to varying levels of familiarity with heat pumps and also to save reading time, the book differentiates between **foundational knowledge** and **expert knowledge**. **Calculations** in the book are always marked with a special symbol and are documented in *italics*. Important terms are highlighted in **bold**.

At the end of primary chapters in Parts I, II, and III, there's a **summarized recap to quickly review key points**.

In this book, frequent reference is made to the **VDI (Association of German Engineers)** as well as various EN standards to provide you with a solid background knowledge. VDI stands for "Verein Deutscher Ingenieure," which translates to the "Association of German Engineers" in English. It is a professional organization in Germany that represents engineers and scientists from various technical disciplines. The VDI develops and publishes technical standards, guidelines, and best practices, aiming to advance technology and promote innovation. They provide a platform for networking, knowledge exchange, and professional development for engineers and scientists in Germany. The VDI's standards and guidelines often play a crucial role in ensuring quality, safety, and efficiency in engineering projects and industries.

Now, dive deep into the world of heat pumps with this book and let the chapter references and other tools guide you on your journey.

Figure 1: Eureka-Luft (EU-L) series by LAMBDA Wärmepumpen GmbH

PART I

«Knowledge means knowing where it is written.»
-Albert Einstein-

1 ENERGETIC FUNDAMENTALS AND TERMINOLOGY

This chapter aims to provide you with the basic principles of energy and thus lay the foundation for the subsequent chapters.

First, let's clarify some terminology. Energy efficiency refers to the measure of energy expended to achieve a certain benefit. In terms of heating technology, this means achieving comfortable room temperatures with the least energy possible.

In many countries and regions, standards exist to ensure energy efficiency in new buildings. While these standards may vary across regions, they all aim to reduce energy consumption and costs, and to minimize environmental impact.

1.1 Types of Energy

Primary energy refers to the natural energy sources that are freely available in our environment. When we discuss primary energy, it can be further divided into two main types:

1. **Fossil Primary Energy:** This energy source comes from non-renewable resources found beneath the Earth's surface, such as oil, natural gas, and coal. These have been formed over millions of years from ancient plants and animals and are a primary source of energy in many parts of the world.

2. **Renewable Primary Energy:** As the name suggests, these are energy sources that are renewable and have a much smaller environmental footprint. They include water (used in hydroelectric power), wind, solar energy from the sun, and geothermal energy from the heat within the Earth.

From primary energy, we derive **End Energy**. This is the energy that's available to consumers after it's been processed and transported. For example, after oil is refined and transported, it becomes heating oil in your tank, or after coal is burned and transformed, the resulting electricity is what you get from your electrical outlet.

Lastly, there's **Useful Energy**, which is the energy directly used by consumers for various purposes. This energy is harnessed when end energy is applied. For instance, when electricity (end energy) powers a light bulb, the resulting illumination is the useful energy.

Figure 2 provides a visual representation of these three types of energy and their relationship.

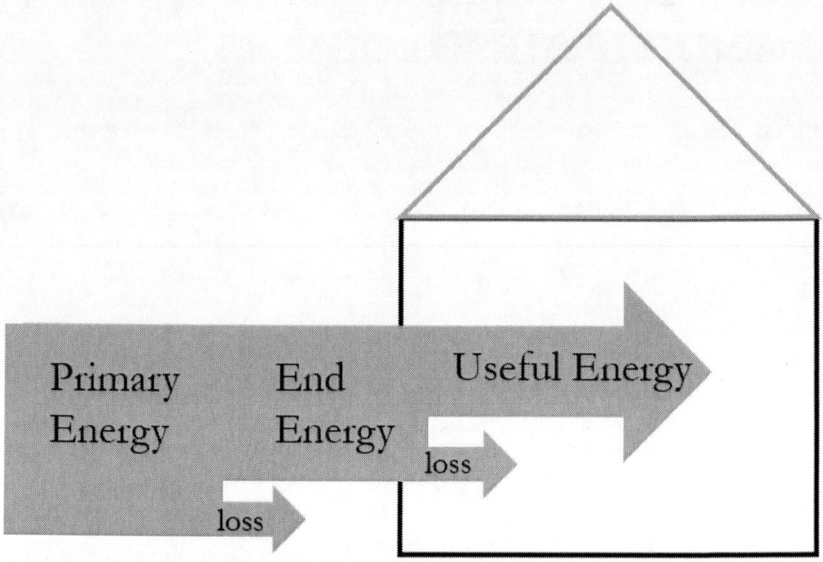

Figure 2: Types of Energy

In many countries, the energy certificate indicates the total energy requirement of a building. The energy efficiency of a building depends on various factors, including the quality of the energy carrier, the efficiency of the heating system, and the building's efficiency.

When looking at different energy carriers, it's interesting to see the impact heat pumps, for instance, have on a household's energy requirement. Utilizing heat pumps can lead to significant energy savings and has the potential to reduce CO^2 emissions. In many countries, the building sector plays a crucial role in efforts to reduce CO^2 emissions.

The amount of fossil primary energy in the useful energy is known as the **primary energy factor**. The subsequent Table 1 provides examples of primary energy factors for various energy carriers.

Comparing various heating technologies can be achieved by categorizing different space heating devices, combined heating devices, and water heaters according to their energy efficiency classes. This is done using **seasonally adjusted space heating energy efficiency (ETAs or annual heating efficiency)** or **water heating energy** efficiency metrics. By calculating the ratio of the covered heating demand to the annual energy demand, the two parameters, **seasonally adjusted space heating energy efficiency** or water heating energy efficiency, can be determined.

Energy sources	"Primary Energy Factors Non-Renewable Share"
Displacement flow mix for CHP	2.80
grid-based	1.80
Heating oil, natural gas, liquefied petroleum gas, hard coal, biogas, bio-oil	1.10
Lignite	1.20
Wood	0.20
locally generated (from photovoltaic or wind power), geothermal, solar thermal, ambient heat, ground cooling, ambient cooling, waste heat	0.00

Table 1: Examples of primary energy factors

It's essential to account for a variety of factors throughout the year. The general assumption is that a heating system operates at partial load 85% of the time during transitional periods and at full load 15% of the time during winter, providing a realistic evaluation of the system. To compare the various technologies of different heating devices, they are classified into specific **energy efficiency classes**.

The efficiency class of a particular device can be determined from its energy label. The efficiency ratings of various water heaters and heating devices are associated with a standard efficiency scale as found on international energy labels.

1.2 Efficiency Calculation of a Heat Pump

To calculate the efficiency of a heat pump, depending on the specific type of heat pump, it's necessary to determine either the **Coefficient of Performance (COP)** or the **(Heating) Seasonal Performance Factor (SPF or Heating Seasonal Performance Factor = HSPF)**. While in some regions and up until specific past dates, achieving certain efficiency benchmarks might have qualified for governmental incentives, as of more recent standards, the SPF (HSPF) serves primarily as a theoretical efficiency analysis for a prospective heat pump installation. Meeting specific SPF (HSPF) benchmarks is no longer a direct prerequisite for obtaining certain incentives; instead, a listing on approved heat pump databases with demonstrated efficiency suffices in many jurisdictions. The calculation of the SPF (HSPF) can typically be achieved using manufacturers' software or through relevant governmental or industry websites.

The **Coefficient of Performance (COP)** represents the ratio of the useful heat output of an electric heat pump during heating operation at specific operating conditions (at a given time), compared to the electrical power used to drive the compressor, auxiliary drives, and control system. Thus, the COP is the ratio of the heating capacity to the electrical power consumption at a particular operating point under defined temperature conditions. It provides a means of comparing the efficiency of heat pumps. The COP is determined on a test bench according to DIN EN 14511 and then plays a role in the calculation of the annual performance factor. It's crucial to understand that the COP focuses solely on the heat pump itself and not the entire heating system. The efficiency of the entire heating system is assessed using the (Heating) Seasonal Performance Factor (SPF/ HSPF) discussed below.

The **(Heating) Seasonal Performance Factor (SPF / HSPF)** refers to the entire year and not just a specific operating point under defined temperature conditions, as with the COP. When discussing the SPF, it's essential to differentiate between the **calculated Seasonal Performance Factor (SCOP = seasonal coefficient of performance)** and the **measured Seasonal Performance Factor (SPF / HSPF)**. Figure 3 illustrates the differences between the COP and SPF spread out over the year.

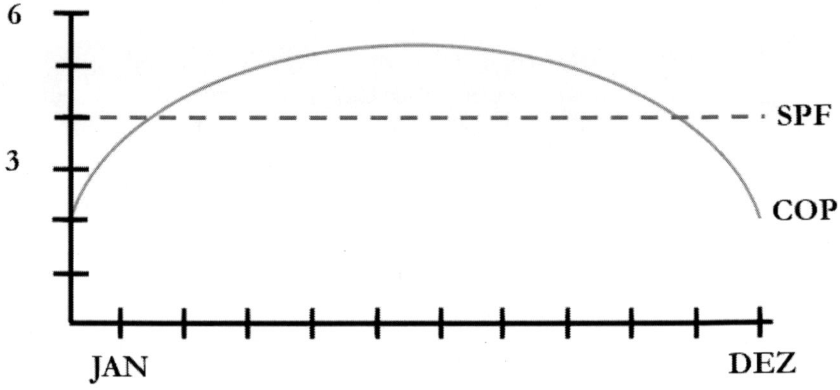

Figure 3: Illustration of COP and SPF (HSPF) throughout the year

The **SCOP (Seasonal Coefficient of Performance)** describes the calculated ratio of the useful heat output by an electric heat pump over the year in relation to the electrical energy used to drive the compressor, auxiliary drives, and control system. The **SPF (Seasonal Performance Factor; HSPF)** describes the ratio of the useful heat output by an electric heat pump over the year, determined using measuring devices, relative to the likewise measured electrical energy used to power the compressor, auxiliary drives, and control system.

To calculate the (Heating) Seasonal Performance Factor of a specific heat pump system, the VDI 4650 Part 1 is used. The (Heating) Seasonal Performance Factor varies depending on the heat source available to the heat pump (refer to Chapter 4), with the Coefficient of Performance (COP) being included in the calculation. For instance, if a property owner utilizes exhaust air as a heat source, the heat pump typically operates with a mixture of outside air. This scenario implies that a simplified calculation of the (Heating) Seasonal Performance Factor using only VDI 4650 Part 1 is not possible, and the additional application of DIN V 4701-10 is required. It's also crucial to distinguish whether the heat pump is used for space heating or domestic hot water heating.

In this book, VDI 4650, and, where applicable, EN standards, will predominantly be referenced for further calculations. It is emphasized that adherence to these standards and guidelines is critical due to their robustness and relevancy in achieving precise and reliable results.

It's important to note that while VDI standards are germane to our context, American readers might refer to corresponding American standards that provide equivalent guidelines and norms, ensuring the universality and applicability of the discussed concepts, processes, and technologies in different geographical and regulatory contexts. The synthesis of VDI and potentially equivalent American standards ensures the comprehensive applicability and relevance of the discussed topics in a broader, international context, underscoring the importance and practicality of these standards and guidelines.

The exact mode of operation (refer to Chapter 5) also impacts the seasonal performance factor. If the heating demand is not met solely by the heat pump but instead through an electric direct heater (monoenergetic system, see Chapter 5.2) or a secondary source of energy (bivalent system, see Chapter 5.3), a specific determination of the coverage proportion by the heat pump is required. The (Heating) Seasonal Performance Factor can also be calculated with the VDI 4650 Part 1 for systems where the heat pump is supported by solar heating or domestic hot water heating. More unique operations, such as using the heat pump for active or passive cooling (refer to Chapter 3.3), can also have their (Heating) Seasonal Performance Factor determined using the VDI 4650 Part 1. A deeper mathematical presentation of the calculation methods will be omitted here, as it wouldn't be in line with the scope and intent of this book. An illustrative calculation expression is presented in Table 2.

In referencing VDI 4650 and the processes it encapsulates, it is crucial to acknowledge its pivotal role in standardizing heat pump efficiency metrics, which has a recognized counterpart in American standards. This reinforces the universal relevance and applicability of the discussed concepts across different regions and regulatory frameworks.

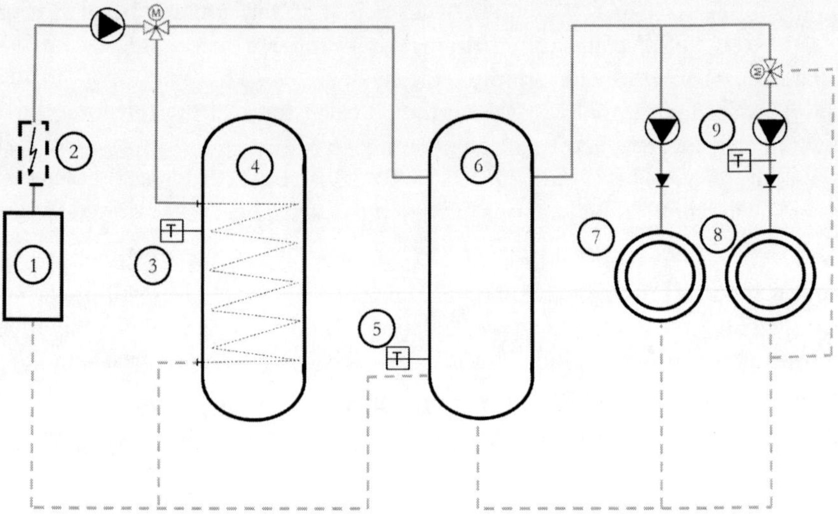

1 Heat pump
2 Electric heating element
3 Temperature sensor for domestic hot water
4 Domestic hot water storage tank
5 Return temperature sensor
6 Heating buffer tank
7 Heating circuit unmixed
8 Heating circuit mixed
9 Temperature sensor for mixer

Figure 4: Boundary limits on WPA and WPHA for calculating the (Heating) Seasonal Performance Factor (SPF/ HSPF)

For comparative analysis, a clear definition of system boundaries in determining the (Heating) Seasonal Performance Factor is necessary. These system boundaries are also referred to as balance limits. In VDI 4650 Part 1, a standardized approach is established that sets the **balance limit for calculating the (Heating) Seasonal Performance Factor of the heat pump system (WPA)** to consider the drive power for the heat source and the secondary heat generator. If one aims to determine the **balance limit for calculating the (Heating) Seasonal Performance Factor for the complete heat pump heating system (WPHA)**, storage losses, pump energy, and heat losses from heat distribution and transfer are also taken into account. Figure 4 illustrates this.

In drawing from VDI 4650, it's important to highlight the uniform standards it sets, which serve as a benchmark that can find parallels in American or other international standards, ensuring global readers understand the context and applicability of these parameters.

To those unaccustomed, the unit-less performance metrics such as the Coefficient of Performance (COP) and the (Heating) Seasonal Performance Factor (SPF / HSPF) might initially seem abstract. **In general, the rule of thumb is: the higher the COP and SPF (HSPF), the better.** A higher value indicates that less electricity is required for heating, which subsequently reduces heating costs. For instance, if one has a COP value of 5, it signifies that the heat pump generates an energy of 5kWh from 1kWh of electricity. This value is a laboratory-measured efficiency metric for the heat pump obtained under ideal test conditions (according to standards) and is rarely achieved in real-world scenarios. The COP specifically pertains only to the heat pump itself, providing a precise metric regarding its performance.

Project information	
Name	xy
Address	xy
Heating setpoint temperature in °F	59
Supply temperature in °F	95
Return temperature in °F	82.4
Information on operating mode and heat source	
Operating mode	monovalent
Standard outside temperature in °F	12.2
Heat source	outside air
Information about the heat pump	
Manufacturer	xy
Model designation	xy
Coefficient of Performance (COP)	3,40
Calculation of annual performance factors (SPF/ HSPF)	
Seasonal Coefficient of Performance (SCOP) of the heat pump in heating mode.	10,9
Seasonal Coefficient of Performance (SCOP) of the heat pump for domestic hot water heating.	10,6
Overall Annual Coefficient of Performance (COP) of the heat pump system.	10,8

Table 2: Part of the exemplary output of a fictitious (Heating) Seasonal Performance Factor (SPF/ HSPF) calculator

Similarly, the SPF (HSPF), which denotes the heat pump's efficiency, suggests with a hypothetical value of 5 that the heat pump has generated a thermal energy of 5kWh from 1kWh of electricity. However, this takes into account several variables such as the insulation of the house, the climate, heating behavior, and even the COP itself. This results in a holistic assessment with the SPF (HSPF) in contrast to the singular focus of the COP. Consequently, while the SPF (HSPF) provides a comprehensive view, it's not strictly suitable for a direct comparison of heat pumps but rather serves as a system comparison. Using a (Heating) Seasonal Performance Factor calculator, one can quickly observe how the SPF (HSPF) varies by adjusting different input parameters. For example, if the insulation of a house is enhanced, the SPF (HSPF) will correspondingly increase.

To provide a global perspective, while the VDI guideline mentioned is a German standard, it's crucial to understand that many countries, including the US, have their respective standards and guidelines. These standards serve similar purposes: ensuring the quality, safety, and efficiency of devices and practices in their respective industries. For a holistic understanding and application, professionals might consider referring to both local and international standards, blending the best practices from each.

With the SCOP, which reflects the efficiency of the heat pump over the course of the heating periods in a year, the seasonal space **heating energy efficiency (ETAs)** can be calculated.

For readers in the US or other international contexts, it's essential to recognize the importance of these metrics. Like the VDI guidelines in Germany, American standards and guidelines serve to ensure the quality and efficiency of devices and practices in their respective industries. Hence, while the SCOP and ETAs are directly referenced from German standards, the core concepts are universally applicable, and similar metrics can be found in various local standards worldwide.

$$\text{ETAs} = \eta\text{s} = \text{SCOP} / \text{primary energy factor} \cdot 100$$

The previously explained primary energy factor (see Chapter 1.1) affects the denominator of the above equation, impacting the space heating energy efficiency. It's evident that the smaller this factor, the higher the space heating energy efficiency will be. Aiming for the highest possible space heating energy efficiency is desirable.

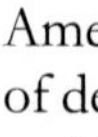

When optimizing the heat pump system, reducing the annual energy consumption is often the top priority. Frequently, a faulty installation or inferior quality of the heat pump is presumed to be the cause. However, often the error lies with the operator. For instance, if one forgets to switch the operation of the heat pump from "heating and hot water operation" in the spring to "hot water operation only" in the summer, it can lead to significant additional energy consumption. Under these conditions, the heat pump would unnecessarily operate several hours daily during the warmer half of the year.

To **measure the (Heating) Seasonal Performance Factor (SPF / HSPF)** and check the efficiency of the heat pump system, a separate electricity meter and a heat quantity meter are needed. The calibration of the separate electricity meter should be observed. Also, it's essential to capture the control system, auxiliary drives, and auxiliary heaters, meaning they should also be connected to the heat pump's electricity meter. It's not uncommon for not all these consumers to be connected to the electricity meter of the heat pump system, leading to a distorted measurement of the seasonal performance factor. While older heat pump models only had a separate electricity meter installed, current models often have an additional heat quantity meter integrated into them. This meter, through sensors in the refrigeration cycle, determines the produced heat amount, displaying the (Heating) Seasonal Performance factor directly on the screen (see Figure 5). If one intends to retrofit a heat quantity meter, an ultrasonic heat quantity meter is recommended due to its lower pressure loss.

Figure 5: Display, heat pump, exemplified

So, how does one determine if the heat pump system operates efficiently? The answer is clear, through the (Heating) Seasonal Performance Factor (SPF / HSPF) and, again, differentiated according to the provided heat sources. The individual types of heat pumps are explained in more detail in Chapter 4.

So far, we have discussed the generation of heat in heat pumps (see also Chapters 3.1 and 3.2). However, in addition to heat generation, heat pumps are often used for cooling purposes (see Chapter 3.3).

When it comes to cooling, we refer to the **Seasonal Energy Efficiency Ratio (SEER)**. The SEER is a metric that assesses the energy efficiency of heat pumps in cooling mode. The SEER value indicates how efficiently heat pumps operate in cooling rooms. It is calculated by the ratio of the total cooling load removed to the electricity consumption required for this purpose over an entire cooling cycle.

A higher SEER value suggests greater energy efficiency, meaning that the heat pump consumes less energy to meet the same cooling demand. This is important for reducing energy consumption, lowering electricity bills, and minimizing environmental impacts.

1.3 Factors Influencing the (Heating) Seasonal Performance Factor (SPF / HSPF)

The factors affecting the (Heating) Seasonal Performance Factor (SPF / HSPF) are diverse:

• **Heat Sink:** The temperature of the heat sink should ideally be as low as possible.

• **Building Envelope:** The thermal transmittance coefficients (U-Value) and heating limit temperatures should be as low as possible.

• **Heat Source:** The abundance of the heat source should ideally be high; the temperature of the heat source should be as high as possible with minimal temperature fluctuations.

• **Coefficient of Performance (COP) of the Heat Pump:** The quality of the heat pump components and their coordination with each other should be of the highest quality and well-optimized.

• **Operator:** The operator behavior/requirements and the setting of the heat pump should ideally be optimal.

• **Weather:** Annual outdoor temperature progression (the influence varies depending on the heat source).

• **Electrical Energy Requirement:** Should ideally be as low as possible.

• **Other Factors:** For instance, support from solar, photovoltaic systems, or a secondary heat generator.

Summary:

Final energy is derived from primary fossil/renewable energy sources (oil, natural gas, water, wind, etc.) and is available to the consumer after deducting transport and conversion losses (such as heating oil in the tank or electricity from the outlet). **Useful energy** is directly utilized by consumers for various purposes and is generated when applying final energy.

The **Coefficient of Performance (COP)** describes the ratio of the useful heat output of an electric heat pump in heating mode, under specific operating conditions (at a given moment), to the electrical power input for driving the compressor, auxiliary drives, and controls. The **(Heating) Seasonal Performance Factor (SPF / HSPF)** refers to the entire year and not, like the COP, to a specific operating point under defined temperature conditions.

The **SEER** describes the energy efficiency of the heat pump in cooling mode.

2 FUNDAMENTALS OF HEAT PUMP SYSTEMS

Here, you will learn the technical fundamentals of a heat pump system in its various configurations.

2.1 Structure of the Heat Pump System

A heat pump system primarily consists of three components: the **heat source system (HSS)**, the **heat pump (HP)**, and the **heat usage system (HUS)**. The heat source system extracts the necessary energy from the environment, which the heat pump then utilizes by adding as little drive energy as possible. The heat usage system stores or distributes the usable heat energy within the house (see Figure 6). In this chapter, we will delve into the structure and function of heat pumps.

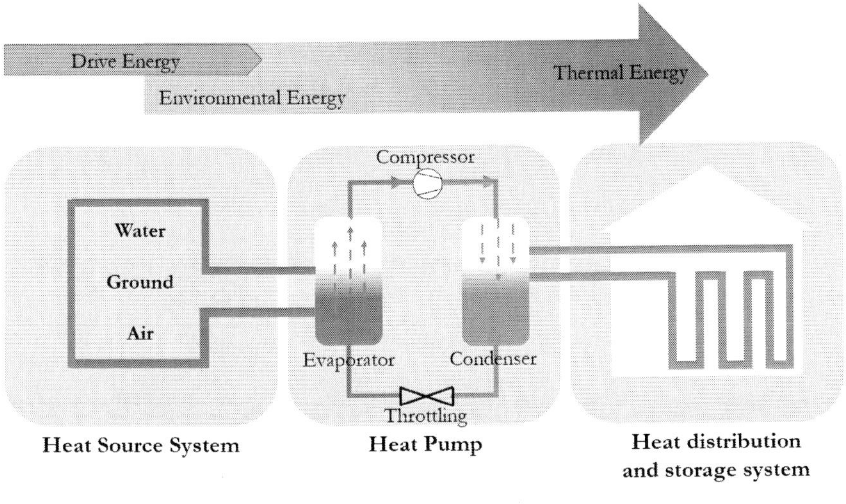

Figure 6: Principle of Operation: Heat Pump

2.2 Operating Principles of a Heat Pump

Two distinct operating principles are identified for heat pumps:

- Compression heat pump
- Sorption heat pump

The compression heat pump is the most widely used operating principle for heat pumps today. While in the compression heat pump the compression of the refrigerant, and the accompanying heating, occurs in a mechanical compressor, in the sorption heat pump, the compression takes place thermally through desorption. A detailed explanation of the functioning of both principles can be found in Section 2.3.

2.3 Structure and Function of a Heat Pump

2.3.1 Structure and Function of a Compression Heat Pump

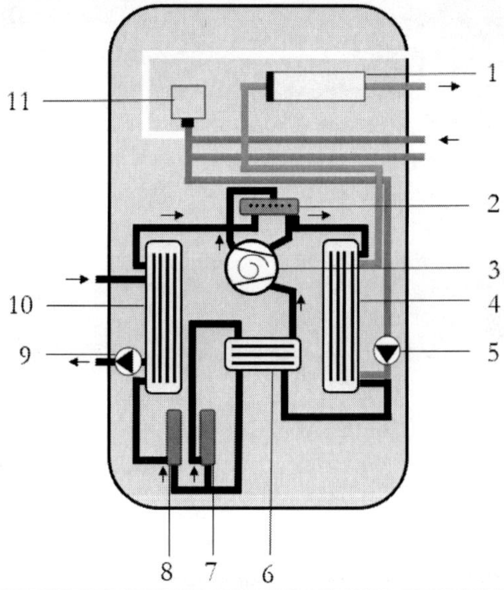

1	Electric auxiliary heating	7	Expansion valve, additional evaporator
2	4-way reversing valve	8	Expansion valve
3	Scroll compressor	9	Brine circulation pump
4	Condenser	10	Evaporator
5	Heating pump	11	Domestic hot water switching valve
6	Additional evaporator		

Figure 7: Structure of a Heat Pump

The function and structure of a compression heat pump, as depicted in Figure 19, aligns with a thermodynamic cycle. Using thermodynamic cycles, energy transformations can be realized with the heat transfer medium (refrigerant). This is represented in Figure 7. The cycle of the heat pump can be likened to that of a standard refrigerator. A refrigerator also has the three main components of a heat pump. The distinction between the two appliances lies in their purpose. While the refrigerator aims to cool the contents stored within, the primary objective of the heat pump is typically the opposite, which is to heat.

The heat pump process within the device is also termed as the refrigeration cycle and can be delineated in **four stages** with **four fundamental components**.

Stages: Heat Input → Work Input → Heat Output → Work Output

Fundamental Components: Evaporation → Compression → Condensation → Expansion

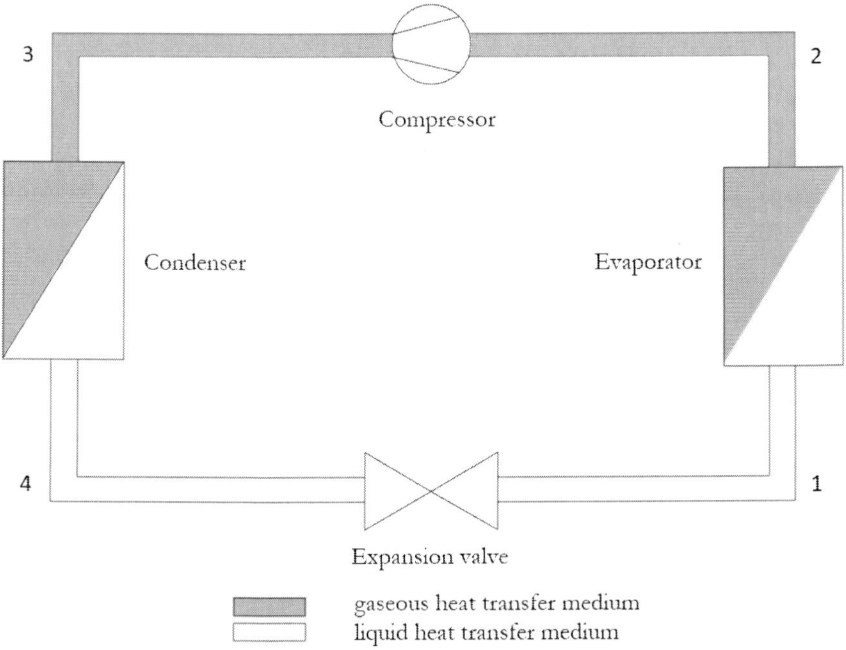

Figure 8: Thermodynamic Cycle of a Compression Heat Pump

The **first stage** of the process occurs in the **evaporator (heat exchanger)** and describes the **intake of heat** (see Figure 8). Heat energy is sourced from the heat source and transported through the evaporator. The still-liquid heat transfer medium (refrigerant) undergoes warming due to this heat energy until it reaches its low boiling point, begins to boil, then condenses, and ultimately exists as a gas. Depending on the heat source medium, the design of the heat exchangers can differ. For air as a heat source, it's often aluminum-copper finned heat exchangers, while for water as a heat source, stainless steel plate heat exchangers are common. In cases of more aggressive water, titanium can be used instead of stainless steel.

In the **second stage**, the process takes place in the **compressor** and is defined as the **input of work** (see Figure 8). The compressor compresses the gaseous heat transfer medium. This stage involves the addition of compression work.

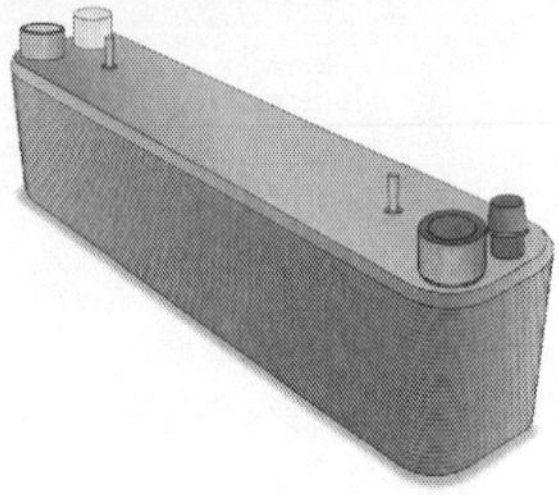

Figure 9: Evaporator, exemplary

In the **second stage**, the process takes place in the **compressor** and is defined as the **input of work** (see Figure 8). With the compressor, there is a compression of the gaseous heat transfer medium. This involves adding compression work. As a result, the temperature of the gaseous heat transfer medium rises. Most of the work energy is converted into compression work.

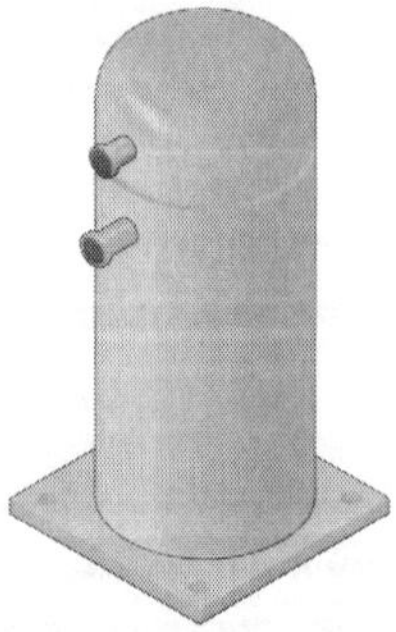

Figure 10: Compressor, exemplary

Here, hermetically sealed scroll compressors, reciprocating or rotary compressors are used. Rotary compressors are generally used in lower performance ranges.

Since reciprocating compressors are sensitive to liquid shocks, scroll compressors, which are less sensitive to droplets in the refrigeration cycle, are often used. The scroll compressor, with its fixed spiral, operates more quietly and, with fewer wear parts, has a longer lifespan (40,000 – 60,000 operating hours). This explains why heat pump manufacturers typically offer a 10-year warranty on the scroll compressor, while only offering a 5-year warranty on the entire heat pump cycle. Reciprocating compressors are usually used in hot water heat pumps and specific applications.

Compressors are available with a fixed speed or inverter technology. For air as a heat source, inverter technology has mostly become the standard. However, inverter technology will increasingly be represented in other heat pumps in the future (see Chapter 7.8 for more details).

During the **third process** stage, the condenser initiates the **release of heat**. The **condenser** condenses the vaporized heat transfer medium, **removing its thermal energy**. During the condensation process, the heating water, which flows at a lower temperature, is passed counter-current to the heat transfer medium through a plate heat exchanger. The heat transfer medium cools down during this process and condenses. The extracted energy enters the heating circuit and can be used for heating purposes.

The **fourth process** stage is called **throttling (expansion valve)** and initiates the **release of work**. The pressure of the now liquefied heat transfer medium is reduced, leading to an expansion of its volume. As the heat transfer medium's volume increases, it cools down, reaching a temperature below the heat source temperature, which describes the starting level. In the conventional refrigerator, this process would be achieved using capillary tubes. Until some time ago, thermostatic expansion valves (TEV) were used in heat pumps for this purpose. Nowadays, electronic expansion valves (EEV) are mostly employed. The heat pump's process is completed and then begins anew.

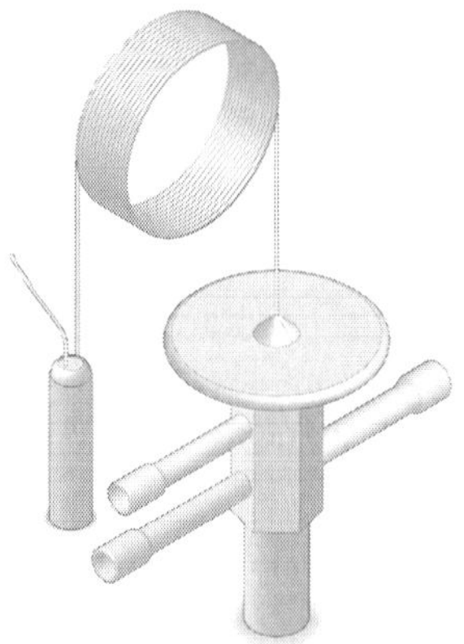

Figure 11: Expansion valve, exemplary

In addition to the essential components of the refrigeration cycle, there are other components which are listed here.

The **filter dryer** is designed to capture moisture and dirt before they enter the expansion valve.

The **sight glass** is used to check for moisture in the refrigeration cycle. Bubbles indicate insufficient refrigerant levels.

If not enough thermal energy can be transferred at the evaporator, the **low-pressure switch** activates.

The **high-pressure switch** activates if not enough produced thermal energy is transferred to the heating surfaces.

To measure the pressure and temperature, **Schrader valves** are used in the refrigeration cycle.

A 4-way reversing valve is significant for heat pumps with the medium of air. With the **4-way reversing valve**, the condenser can be switched to the evaporator and vice versa. This way, for air-to-air heat pumps, the defrost function can be implemented (see Chapter 4.3), utilizing thermal energy from the buffer storage for the defrosting process.

The components of the refrigeration cycle mentioned so far are present in all heat pumps except for direct-evaporation/direct-condensation heat pumps. The only difference is the structure, or the varying arrangement and quality of the components.

Due to the vibrations of the compressor and the high pressures, the refrigeration cycle is subjected to particular stress. The refrigeration cycle must undergo a precise leak test. This should be done before refilling the refrigerant and is based on the environmentally harmful nature of the refrigerant, coupled with the potential for costly repairs.

2.3.2 Structure and Function of a Sorption Heat Pump

The operation and structure of a sorption heat pump follow a reversible physico-chemical cyclic process in which two substances, initially present as a mixture, are separated. Since the cyclic process can proceed in both directions, it's described as reversible. The sorption heat pump is further divided into the **absorption heat pump** (operating with a liquid sorbent) and the **adsorption heat pump** (operating with a solid sorbent). In the following, the functioning will be demonstrated using the absorption heat pump as an example. After the two substances have been separated, they are recombined in the subsequent absorption process, generating heat.

With the reversible physico-chemical cyclic process, energy transformations can be achieved. Even with the sorption heat pump, the cycle's first process stage involves the evaporator, the third stage involves the condenser, and the fourth stage involves the throttle. Thus, the process stages 1, 3, and 4 are similar to those of a compression heat pump. Here as well, the heat pump process within the heat pump can be termed as a **refrigeration cycle** and can be described in **four stages** with **four elementary components**.

Stages: Heat input ➔ Work input ➔ Heat release ➔ Work release

Key Components: Evaporation ➔ Compression ➔ Condensation ➔ Expansion

The primary distinction between a sorption heat pump and a compression heat pump lies in the second stage, namely the compression process. The compression process in the sorption heat pump, unlike in the compression heat pump, is implemented differently and is divided into four sub-processes, which are explained below. Since no traditional compressor is used here, this process is also referred to as **thermal compression**, and the encompassing component is known as the **thermo-compressor**. The other stages can be referred to from the previous section 2.3.1 for the compression heat pump.

The **first subprocess stage in process two** occurs in the **absorber of the thermo-compressor**, describing the **heat release**. In the absorber, the incoming refrigerant vapor is absorbed by a solvent (e.g., water). This results in the release of heat, which is transferred to the heating network via a heat exchanger.

In the **second subprocess stage of process two**, the **solution pump within the thermo-compressor** transports the mixture of substances to the desorber. This requires very little energy and is referred to **as work input**.

The **third subprocess stage in process two** is carried out by the **desorber (thermal compressor) of the thermo-compressor**, causing heat to be added to the mixture of substances (ammonia and water). Based on the different boiling temperatures of the substances, the ammonia evaporates from the mixture. Thus, the gaseous ammonia can be supplied to the condenser for process 3. This action is described as **heat uptake**. The heat energy can be derived from combustion processes using oil or gas or from waste heat utilization.

The **fourth subprocess stage in process** two takes place in the **throttling (expansion valve) of the thermo-compressor**. Through a throttle valve, a solvent is introduced for expansion. During this, volume change work is performed. This process can be described as **work release**.

The advantage of the working principle of a sorption heat pump compared to a compression heat pump lies in the compressor. While the compressor of a compression heat pump often produces significant noise, potentially necessitating soundproofing measures, this is not the case with the thermal compressor of a sorption heat pump. Additionally, the moving parts in a sorption heat pump are considerably fewer, meaning that regular maintenance is less frequent compared to a compression heat pump.

A disadvantage of the sorption heat pump is the generally higher procurement costs due to the variety of components.

One drawback of the sorption heat pump is typically the higher procurement costs due to the diversity of components.

2.4 Gas-Powered Heat Pump, an Alternative for Drive Energy

As we already know, heat pumps utilize the free renewable primary energy or environmental energy from groundwater, earth, or air to generate heat (see Chapter 4). However, drive energy is also needed, which is typically electricity. But the drive energy doesn't necessarily have to be electricity; gas can also be used in this regard. With the gas-powered heat pump, the heat sourced from the environment is compressed using a gas burner. This burner could also operate on biogas, which would again be a renewable energy source. Typically, with gas-powered heat pumps, higher flow temperatures can be achieved with less environmental heat. Due to these higher flow temperatures, gas-powered heat pumps are suitable for older, poorly insulated houses.

Since electric heat pumps are by far the most commonly used, the rest of the book will focus exclusively on electrically powered heat pumps.

Summary:

A heat pump system primarily consists of three components: the **heat source system (HSS), the heat pump (HP)**, and the **heat utilization system (HUS)**. The compression heat pump is currently the most widely used working principle of a heat pump. The process within the compression heat pump is also referred to as a refrigeration cycle and can be described in four stages with four fundamental components. Apart from the electrically powered heat pump, there's also a gas-powered version, with electric heat pumps being far more prevalent in application. Therefore, the rest of the book will focus exclusively on electrically powered heat pumps.

3 APPLICATION AREAS OF HEAT PUMP SYSTEMS

Here, you will learn about the key application areas of heat pump systems.

3.1 Heat Pump Systems for Heating

Heating a building is by far the most significant application when using a heat pump and is thus the primary focus of this book. The produced heat quantity is transferred to heat the building or its structure. In this context, there are two different methods to heat a building. The **first method** involves heating the building using water warmed by the heat pump in a heating circuit via surface heating systems (underfloor heating, etc.) or radiators to warm the rooms of the building. This is achieved, for example, through brine/water, water/water, or air/water heat pumps. It should be noted beforehand that the first "medium-word" in the heat pump designation (e.g., air-water heat pump) corresponds to the medium that transfers the heat to the heat pump. Consequently, the "second" "medium-word" in the heat pump designation (e.g., air-water heat pump) pertains to the medium that releases the heat in the building or the rooms. Thus, it becomes clear that all the heat pumps mentioned here transfer heat to the rooms in the building using the medium of water (known as "heating water"). Further details about the heat sources and the various types of heat pumps are explained in Chapter 4.

For the **second method**, heat is transferred to the rooms via air, not using heating water. So, the heat pump directly warms the rooms through the room air. Such a heat pump would be an air/air heat pump, and the previous clarification regarding the "medium-words" applies here as well.

3.2 Heat Pump Systems for Domestic Hot Water Heating

In addition to heating, the heat pump system can also be used for domestic hot water heating (potable water heating). It should be noted beforehand that due to the technologically higher feed temperature required, using heat pump systems for hot water preparation has a lower efficiency compared to the sole use of the heat pump system for heating. The greater the proportion of hot water preparation in a building, the lower the overall (Heating) Seasonal Performance Factor for the heat pump system will be. Thus, it becomes clear that in a house with an increasing number of occupants, while the demand for drinking water increases, the heating requirement does not, yet the efficiency of the heat pump system decreases.

3.3 Heat Pump Systems for Cooling

Just as the heat pump is used for heating or domestic hot water heating, it can also be utilized for the opposite purpose, cooling. But how is this implemented? Firstly, it's essential to differentiate between the **types of cold transfer in passive (static)** and **circulated air (dynamic) cooling** by heat pumps. Furthermore, the cooling **technique needs to be categorized** into **active** and **passive cooling**. Therefore, there's either passive or circulated air cooling, and within these, either active or passive cooling.

In **passive cooling**, cooling is achieved through cooling surfaces, which are the actual heating surfaces during heating. The cooling occurs through the so-called "cold falling air." Warmer air rises due to its lower density compared to cold air, generally moving from bottom to top. This means that, assuming no disproportionate air circulations occur, it is always a few degrees warmer near the ceiling. Consequently, it's cooler at the floor level, which explains the term "cold falling air." The previous explanation also clarifies that for the surface heating used in passive cooling, ceiling heating for cooling is more efficient than underfloor heating. Blower convectors can also be used for cooling. However, when planning for them, it's essential to account for an electrical connection and a condensate drain. Radiators are not suitable for room cooling because their relatively small surface area, compared to the room volume, allows for limited heat transfer and favors condensation. It's worth noting here that if you intend to use your surface heaters for both heating and cooling, it's crucial to avoid condensation on the cooling surfaces. This is ensured by preventing the room air's dew point from being reached, thus avoiding the condensation of water on the cooling surfaces. For this purpose, a room temperature and humidity sensor, connected to the heat pump controller, is installed, ensuring that the supply temperature remains at least 3-5 degrees Fahrenheit, depending on the type of floor, above the dew point temperature. Furthermore, it's essential to use diffusion-tight insulation for the pipes and fittings. When using underfloor heating, it's necessary to check if the materials can be used for both heating and cooling. The thermostats in such configurations should always be switchable between heating and cooling modes. If room humidity rises, the cooling capacity should be limited due to the risk of reaching the dew point.

As the name suggests, **circulated air cooling** occurs through air currents, akin to an air conditioning system.

Now, let's discuss active and passive cooling.

Heat pumps suitable for **active cooling** are also referred to as reversible heat pumps. They use a 4-way valve in the refrigeration cycle to reverse the circuit. This turns the evaporator into a condenser that transfers the room heat to the refrigerant via the surface heating, and the condenser into an evaporator.

With the reversible heat pump, the existing temperature level is "actively" cooled down using the compressor performance of the heat pump. The detailed operation of the heat pump can be found in Chapter 2. It's essential to know that if the heat pump is also used for domestic hot water heating, on days when room cooling is required, both the cooling operation and continued domestic hot water heating are maintained with this pump. These are opposing processes within the heat pump.

Passive cooling is a fairly straightforward principle (see Figure 12). On warm summer days, for example, if room temperatures are higher than those of the ground (using a brine/water heat pump to harness geothermal energy (see Chapter 4.2)), the heat in the house can be absorbed via the surface heating system and discharged back into the ground through the return flow. It's important to note that when using geothermal energy, flat collectors are not suitable for cooling with heat pumps. When using water as a heat source (see Chapter 4.1), legal water regulations must be adhered to during cooling. This process is also referred to as "natural cooling." So, while during heating the geothermal heat is extracted, heated by the compressor, and then returned to the ground cooler after balancing temperatures in the rooms to be heated, the compression process of the geothermal heat is omitted during cooling, with room heat being extracted and returned to the ground (see Figure 12). This is ensured in smaller heat pump systems via an integrated intermediate heat exchanger and a 3-way switching valve. In larger heat pump systems, ready-made assemblies are installed next to the heat pump for this purpose. Therefore, during cooling, only the brine and heating circulation pumps run, and the compressor remains off. If a prolonged heatwave occurs, it's possible that the ground could warm up from the heat discharge to such an extent that it reduces the cooling. In such cases, active cooling solutions, such as air conditioners or chilled water units, would provide relief.

By continuously introducing heat into the ground, so-called "heat bubbles" can form, which can maintain a heightened temperature level in the ground well into the colder season. This, in turn, provides more geothermal energy at the start of the heating period, thereby increasing overall efficiency. Passive cooling can also be applied as **controlled ventilation for living spaces**. In the winter, pre-heating to outside air can take place via an intermediate heat exchanger from the ground probe, replacing the need for electrical auxiliary heating. During the warm season, summer, the temperature of the incoming air can be reduced with passive cooling.

Active cooling can be implemented as both silent cooling and circulated-air cooling. Passive cooling is only implemented as silent cooling, which excludes air-to-air heat pumps. Generally, passive cooling is preferable to active cooling, as the energy consumption for active cooling is typically higher.

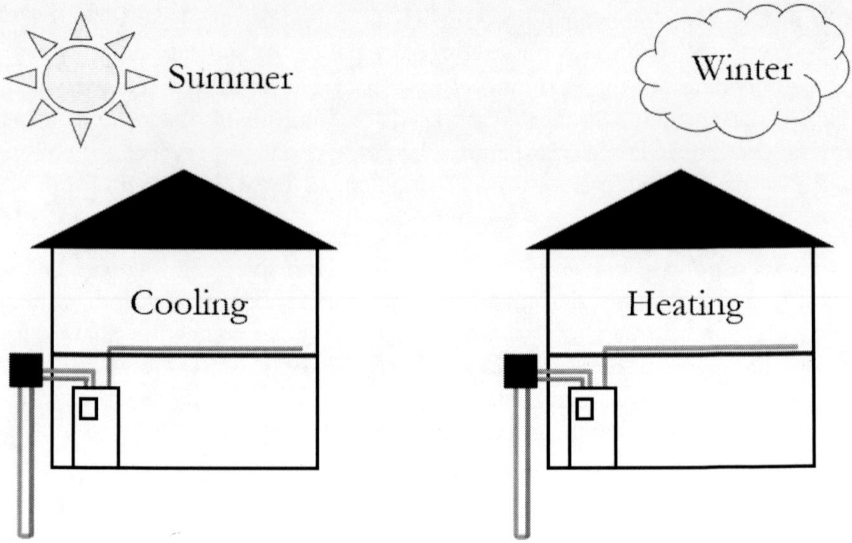

Figure 12: Functional principle cooling and heating

Summary:

The heat pump can be used for **heating**, domestic hot **water heating**, as well as for **cooling**.

4 HEAT SOURCES AND RESULTING TYPES OF HEAT PUMPS

In general, there are three fundamental heat sources:

- Water (Groundwater)
- Earth (Brine)
- Air

Depending on the heat source, there are specifically designed heat pumps for each. Figure 13 illustrates the relationship of the three heat sources to the efficiency of the heat pumps. It becomes evident that the efficiency of the groundwater heat source is the highest. However, its availability is limited. Conversely, with the air heat source, while there is almost unlimited availability, its efficiency is the lowest due to fluctuating temperatures.

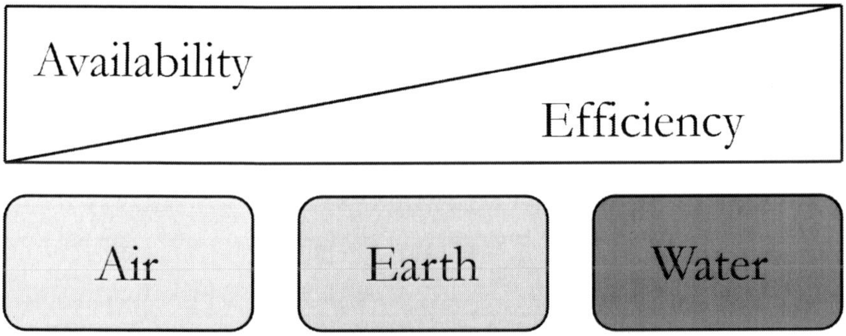

Figure 13: Availability-to-efficiency ratio

This chapter consists of four primary subsections. In the first three subsections, the heat sources of water, earth, and air, as well as their associated heat pumps (HP), are explained. In the naming of the heat pumps, the mediums are usually mentioned twice (e.g., air/water heat pump). The first medium mentioned refers to the heat source (water, earth, air), i.e., from where the heat is derived, and the second medium refers to the medium through which the heat energy is released in the building (e.g., water in the heating circuit or the heat is directly delivered to the building through warmed air). In the fourth subsection, special designs, both unique types of heat pumps and special designs of heat sources, are discussed.

Appendix C features a matrix of the most important heat pumps explained in this chapter, listed in a **heat pump matrix**, to assist you in choosing your heat pump.

4.1 Heat Source: Water

Let's now turn our attention to groundwater, which, from an efficiency perspective, is the best heat source. Using groundwater as a heat source for the heat pump requires the construction of a well system. A consistent water temperature of 44-50°F can be expected from a depth of about 2m onwards. To utilize groundwater as a heat source, a sufficient groundwater level at the mentioned depth is necessary. As a rule of thumb, approximately 250 l/h of water is required for 1kW of heating power. With groundwater as the heat source, the heat pump can guarantee excellent performance figures throughout the year. Such a facility requires approval from the local water authority at the responsible district office since the user interferes with the groundwater level.

The construction of a groundwater heat source is illustrated in Figure 14 and Figure 15. Two wells are necessary for this system. The groundwater is drawn from the first well (extraction well or also withdrawal well) and directed to the evaporator of the heat pump. The energy necessary for the cycle is then extracted from the evaporator of the heat pump, and subsequently, the water is returned to the second well (infiltration well or also disposal well) with a cooling of up to 41°F. Due to the consistently high groundwater temperature of 44-50°F mentioned earlier, the groundwater can be used directly.

When constructing the well system, mutual interference between the extraction and infiltration wells, as well as an "energetic short circuit", should be avoided. Therefore, a minimum distance of at least 15m between the two wells should be maintained.

It is also important to consider the direction of groundwater flow (see Figure 14), which may require extensive hydrological investigations. This alone indicates that a property of an appropriate size is required for this purpose. A certified well construction company with sufficient experience should be used to set up the well system.

The system consists of two well shafts. The shaft of the extraction well is equipped with a filter pipe surrounded by filter gravel. The filter gravel is meant to purify the water before extraction, thus preventing contaminants in the cycle. The submersible pump is placed inside the extraction well. To protect the submersible pump from running dry, it is usually surrounded by a float switch. The heat exchanger ensures that the groundwater releases its energy to the heat transfer fluid, thereby lowering the temperature of the groundwater. Different types of heat exchangers, such as tube bundle heat exchangers, spiral tube heat exchangers, or plate heat exchangers, can be used.

1 Heat pump
2 Buffer tank
3 Domestic hot water tank
4 Radiator
5 Underfloor heating
6 Hot water pipe
7 Extraction well
8 Injection well
9 Groundwater flow direction

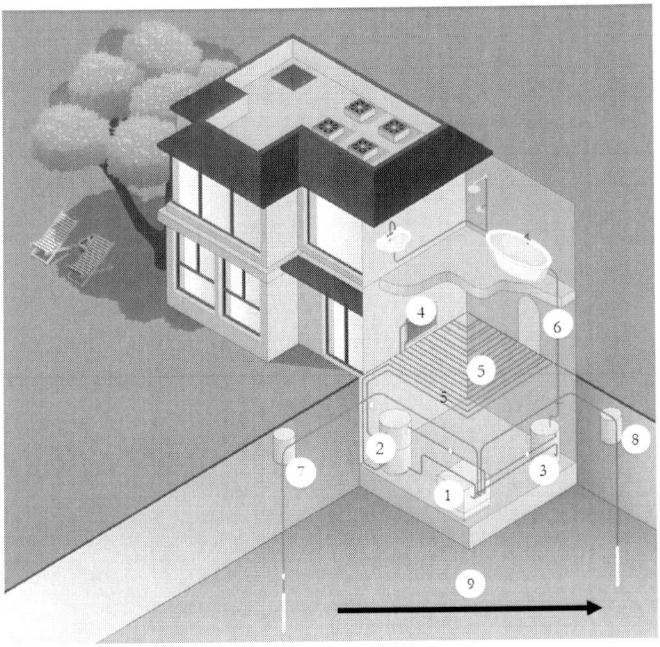

Figure 14: Structure of groundwater heat source

Before entering the heat exchanger, the groundwater, which has already been coarsely filtered by the gravel in the extraction well, is further purified to prevent contamination of the heat pump circuit. If there is still no guarantee of persistent contaminants, iron loads, manganese loads, and suspended solids, an intermediate heat exchanger should be integrated into the system. This intermediate heat exchanger can be cleaned, but it is again associated with losses. This requires a detailed investigation. A water meter is installed in the system to calculate the exact extraction quantity of the extraction well. Plastic pipes are typically used for water transport to the heat pump, but more specialized pipe materials can also be used.

As mentioned earlier, the constant heat source makes it possible to feed the water extracted from the extraction well directly to the heat exchanger. This allows the heat pump to be installed as a compact unit in the house, saving space.

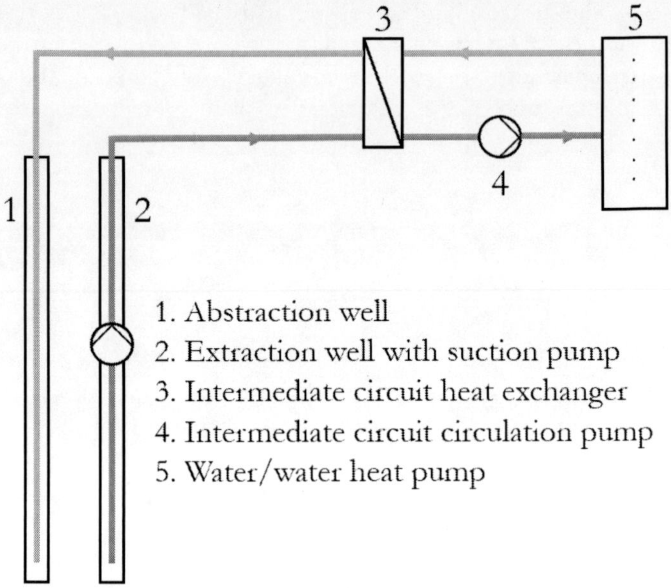

Figure 15: System Groundwater Heat Source

4.1.1 Heat Source Water: Water/Water Heat Pump

Although the water/water heat pump (also known as the groundwater heat pump) is among the most efficient types of heat pumps, such a system also comes with numerous requirements that can be costly without sufficient expertise. Therefore, it is strongly recommended to consult with the appropriate professionals and not act hastily. In some circumstances, it might be advantageous to consider a brine/water heat pump with an earth probe instead of a water/water heat pump. For example, a water/water heat pump might be reconsidered in favor of a brine/water heat pump with an earth probe if the submersible pump has to hang very deep due to hydrological conditions. In the following chapter, we will focus on the heat source of the ground.

4.2 Heat Source: Ground

Heat pumps with ground probes were actually the origin of all heat pumps and were predominantly used at the beginning of this technology. The reason for this was a certain skepticism towards the other two heat sources, water and air, and especially with air-based heat pumps, there were comparatively high losses at that time. In our latitudes, geothermal energy can provide the necessary energy even on extremely cold winter days, so that the heat pump can operate economically. The ground to be used as a heat source heats up mostly by the sun up to a drilling depth of 10 meters. If you drill deeper, the seasonal differences become increasingly smaller. Figure 16 shows the seasonal and drilling depth-dependent temperature curves of the ground.

The effect of sunrays hitting an obstacle can be divided into three types of reactions: absorption, reflection, or letting the sunrays pass through. The reaction of the sun to the ground can largely be described as absorption. A much smaller part here, dcpcnding on the surface of the earth, is reflection. Due to the absorption, the ground heats up, whereby the earth mass is proportional to the heat source. So if you increase the mass of the earth, the amount of heat that can be absorbed by the ground also increases. Thus, grounds with high mass are a good heat storage. Using the ground as a heat source, in turn, causes a heat extraction from the ground. To ensure continuous use of the ground as a heat source, it is necessary for an energetic regeneration of the used area to take place. To meet the requirements of a sufficiently high and constant amount of heat as well as the necessary regeneration, the following things should be considered. The planting of the used area should be kept as low as possible so that rainwater is available to the ground. Good soil irrigation should be available, and from an energy perspective, additional rainwater irrigation is recommended.

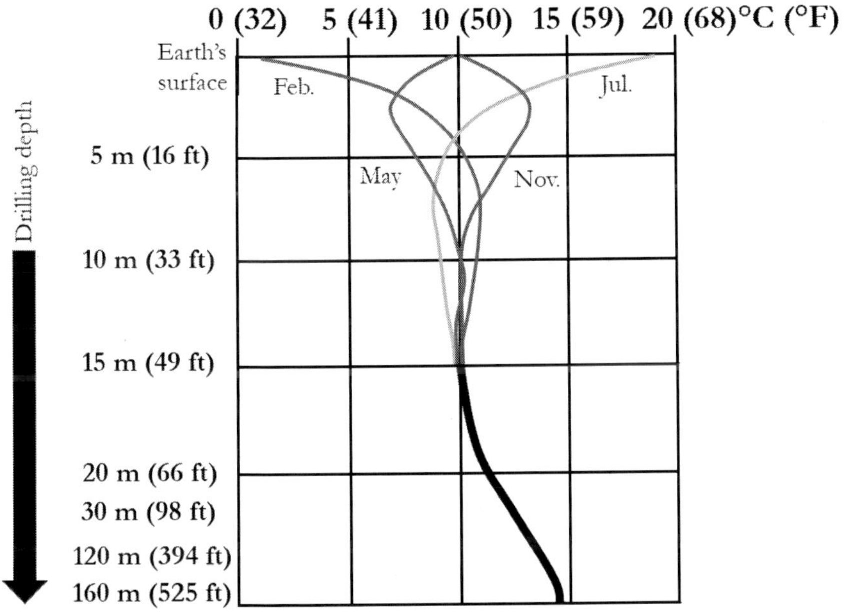

Figure 16: Seasonal and borehole depth-dependent temperature variations of the ground

The usable area should be as free from shading as possible throughout the day.

Basically, there are two different methods to utilize geothermal energy. The first method is geothermal probes (ground probes) that go deep into the earth to draw geothermal energy. The second method describes ground collectors, which are laid horizontally in the ground rather than going deep. Both methods have their justification and need to be weighed up depending on the individual situation. For example, with a small plot size, there may be no other option than to use ground probes when utilizing geothermal energy. Additionally, there are other designs for drawing energy from the ground that will also be discussed.

Geothermal energy itself can be sourced in two different ways for both mentioned methods (geothermal probes and ground collectors): using brine (Sole) or direct heating. With the brine, a separate circuit separated by the intermediate heat exchanger is installed. In this circuit, the heat transfer medium brine circulates to absorb geothermal heat. Brine is a circuit where an anti-freezing agent is added to achieve a freezing point below that of water, thus preventing freezing. The brine is transported in the ground by a brine circulation pump. Due to the two circuits existing through the intermediate heat exchanger, additional losses with the brine variant are inevitable. The second variant, direct heating, causes the working medium of the heat pump to circulate directly as a heat transfer medium in the geothermal probe or ground collector. With this variant, additional losses due to a second circuit are eliminated.

4.2.1 Borehole Heat Exchangers (Vertical)

In the following, the borehole heat exchanger variant is discussed in more depth. The borehole heat exchanger goes into the depth between 50 and 250m (164 and 820ft) of the ground and uses it as a heat source. Figure 17 shows a system based on ground source probes. A consistent year-round temperature of 50°F can be expected from a drilling depth of 15m (49ft) to 50m (164ft)). If drilled deeper from 50m, the temperature rises. An increase of 3F/100m (328ft) can be assumed. This means that the borehole heat exchanger can absorb more energy in the circulating circuit as the depth increases. As previously mentioned, the relatively small horizontal space requirement of the ground source probes is a decisive advantage.

Once the drilling (see Figure 19) is completed, a probe consisting of 4 tubes with a diameter of approximately 32mm (1.26in) and made of polyethylene (PE) is introduced into the borehole (see Figure 18). Typically, U-tubes are interconnected, forming double U-tube probes or a bundle of tubes held together with clamps. They are referred to as supply and return lines. In the center of the bundle is the injection tube (pipe) (see Figure 19). Once the bundles are positioned in the borehole, a bentonite-cement mixture is pumped in through the injection tube.

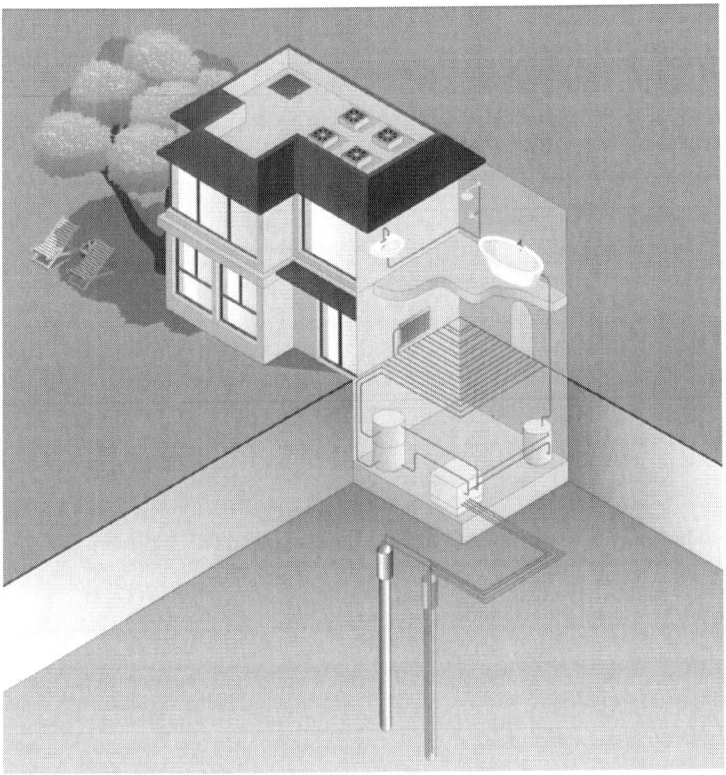

Figure 17: Ground Source Heat Pump System

This bentonite-cement mixture, filled from the bottom up, presses the bundle into the earth's heat source, improving heat transfer from the earth to the bundle and increasing extraction efficiency. Another reason for the pressing is to protect deep groundwater layers from surface water flowing down through the borehole. Rarely, simple U-tube probes or coaxial probes are used, with coaxial probes being used in areas with depth restrictions and as storage probes. Mixing groundwater from different layers is not permitted. The size of the space between the soil and the U-tubes is crucial for extraction efficiency. A larger area produces more extraction power. To increase extraction power, it is entirely possible to create multiple bundles and associated boreholes, which are then combined through distributors and collectors. Typically, a minimum borehole spacing of 5m should be maintained.

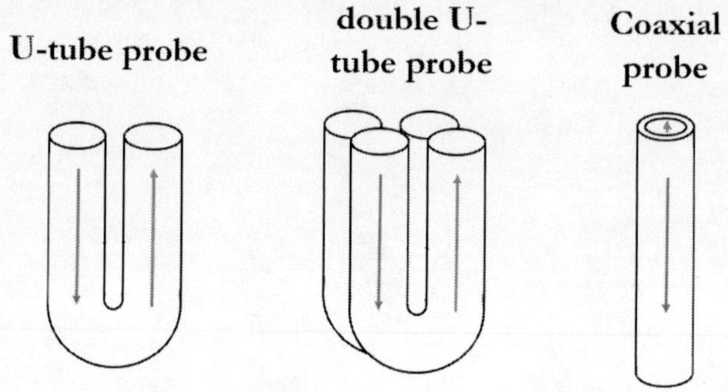

Figure 18: Different ground loops

The number and respective depth of the geothermal probes depend on the energy requirements of the building and the yield of the respective soil. The required extraction capacity should be known before work begins. This has the advantage that a minimum extraction capacity can be contractually agreed upon with the drilling company at a fixed price. The approval process, assessment of geological conditions, drilling depth, number of boreholes, test drillings, and any misdrillings are thus included in the fixed price and are carried out by the drilling company.

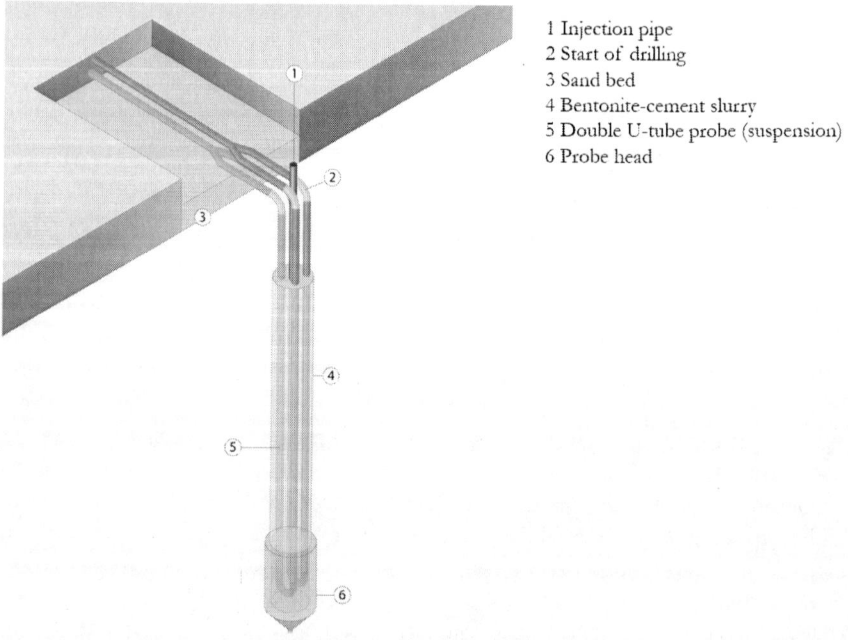

Figure 19: Borehole, exemplary

The customer therefore does not bear the risk if these services are contractually assigned to the drilling company.

Typically, for single or two-family homes, two geothermal probes with a drilling depth of up to 120m are planned. The diameter of the drill is usually 152mm, and depending on the situation, flushing drillings or the down-the-hole hammer method are used. During the drilling itself, if not previously determined geologically, it can happen that an artesian well is drilled into (see Figure 20). This refers to pressurized groundwater where the hydraulic potential is so high that the groundwater rises above the earth's surface without the use of pumps. If an artesian well is drilled, water can leak out of the borehole, or even shoot out in the form of a meter-high fountain with considerable pressure. In such a case, precautions should be taken beforehand to ensure that the escaping groundwater can flow away in a controlled manner and does not flood any buildings nearby. But another phenomenon can also occur. After the groundwater, the artesian well, has released its pressure, there can be subsidence in the underground, which in turn can affect adjacent buildings. These circumstances are not meant to scare or deter from the installation of geothermal probes, but on the contrary, to sensitize that good preparation and a sufficiently insured and experienced drilling company should always be the choice. The company must strictly adhere to the conditions contained in the drilling permit (if only for insurance protection). Furthermore, in areas with artesian pressured water, drilling work must be carried out with increased caution, and precautions should be taken that can unproblematically divert any larger amount of water that may leak from the borehole.

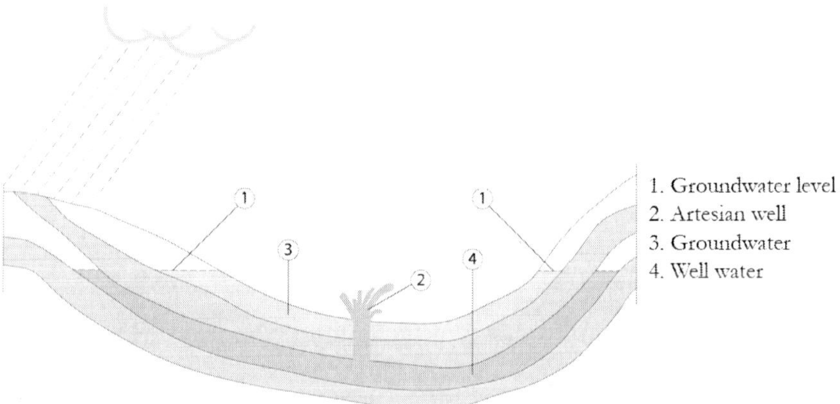

Figure 20: Artesian well

4.2.2 Ground Heat Exchangers (horizontal)

In contrast to the deep (vertical) use of geothermal energy through geothermal probes, the near-surface (horizontal) use through ground heat exchangers can also be an effective means of extracting thermal energy. The surface collector primarily utilizes the earth's energy from rain and sun and only draws a very small portion of the energy from the ground. Typically, plastic pipes of various dimensions are laid horizontally in the ground at intervals of 50 to 80 cm (1.64 to 2.62ft). The installation depth is aimed at between 120 and 150 cm (3.94 to 4.92ft), taking into account the respective frost-free conditions, which guarantee an all-year-round temperature greater than 32°F in the ground. The pipe surface in the soil is a crucial factor for the extraction performance and is proportionally included in the calculation of the thermal output. As a rough guideline, it can be determined that 1.3 to 1.5 times the living space needed for heating must be available as the surface of the ground heat exchangers to provide sufficient heat.

Example Calculation:

If you have a residential building with an area of 200 square meters, you would need a collector area of 30 square meters. The collector area does not describe the surface area calculated from the outer diameter of the collector pipes but rather the total area covered by the pipes. Therefore, collectors measuring 20x15 meters would be sufficient to comfortably heat a 200 square meter living space.

This would amount to a total of approximately 620m of piping. With such a total pipeline length, a significant pressure loss would be associated.

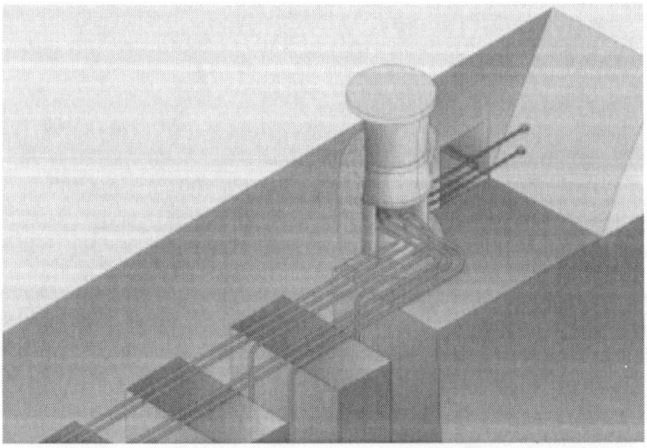

Figure 21: Distribution box, located outside in a shaft

Therefore, the maximum pipe length is usually limited to 100m, and distribution structures are used, which in the given example would distribute 7 collector strands of 89m each in parallel. This ensures that the distribution of the heat transfer medium is consistent, and thus the same pressure loss is present in each distribution strand.

Furthermore, this ensures uniform operation of the distribution strands, and through interposed ball valves, individual strands can be specifically shut off. Such a distributor is recommended to be placed outside the building in a kind of shaft, as this minimizes the number of wall breakthroughs for each line and overall, only requires one supply and one return line through the wall (see Figure 21).

4.2.3 Additional Design Types for Energy Retrieval from the Earth as a Heat Source

4.2.3.1 Ring Trench Collector

To achieve as large a heat transfer surface as possible in a large space without using a geothermal probe, another option is to install the so-called **ring trench collectors** (see Figure 22), also known as ring ditch collectors. Here again, plastic pipes are used, but to achieve the required heat transfer surface in as large a space as possible, they are arranged in a spiral form. This saves space in the ground and can achieve the same extraction performance as with surface collectors. Trench collectors can be laid horizontally, but at shallower depths (about 3m), they can also be laid vertically. They can be arranged in lateral spirals, vertical spirals, two-dimensional or three-dimensional spirals.

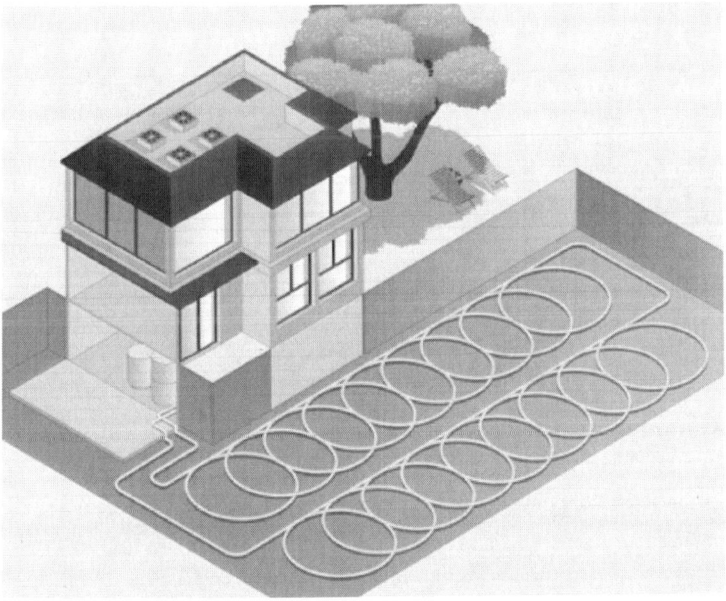

Figure 22: Horizontally laid ring trench collectors

Each ring trench collector is laid in individual trenches, usually parallel to each other. For both horizontal and vertical installation, the distance between the trench collectors is between 3 and 4 meters, with a trench depth of between 2 and 3 meters for horizontal installation. The greater horizontal distance of 3 to 4 meters for the trench collector, compared to the surface collector, is necessary to avoid mutual interference between the trench collectors due to the greater extraction performance in a smaller space and to allow for the natural regeneration of the soil. During installation, the trenches are dug individually, and the excavation from the second trench is used directly to backfill the first trench. This avoids the deposition and interim storage of the excavated material. The extraction performance here also depends on the type of soil, soil moisture content, spiral diameter, diameter of the plastic pipe, the exact material of the plastic pipe, depth, and length. A cost-effective installation compared to geothermal probes is also evident here.

4.2.3.2 Geothermal Baskets

Ground source heat baskets offer advantages similar to the trench collectors, as depicted in Figure 23. Ground source heat baskets can be differentiated into two typical designs. The cylindrical design, which resembles a spiral, and the conical design (basket-like).

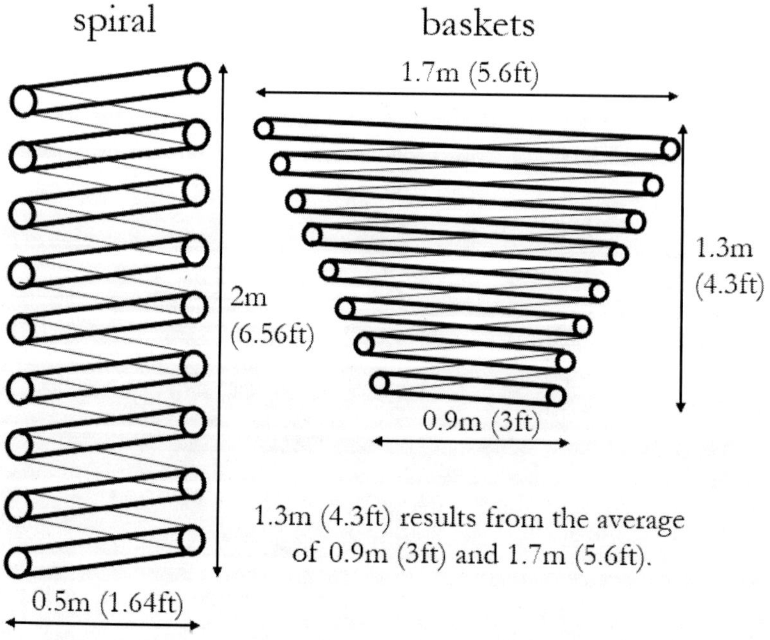

Figure 23: Geothermal baskets

The holes can be dug with an excavator and the geothermal baskets can be installed in them. Since geothermal baskets are prone to uplifting and settling effects, they should be backfilled with sandy or self-compacting fill materials after installation.

4.2.3.3 In-situ Concrete Piles

If, for structural reasons, drilled piles are required instead of a floor slab to transfer the load of a building, this can also be used to install comparatively short geothermal probes made of PE pipes. However, not double-U-tube probes are used here, but only a single U-tube probe is installed.

4.2.3.4 Short Borehole Heat Exchangers

There is also the general option to install short **borehole heat exchangers**. These short borehole heat exchangers would be at a maximum depth of 15m (49ft). Due to their shallower depth compared to classic borehole heat exchangers, these types behave more like horizontal methods, as they are more influenced by weather conditions.

4.2.3.5 Trench Collector

Trench collectors are another method to harness geothermal energy. According to Figure 24, trench collectors are positioned on a trench wall that is up to 3m (9.8ft) deep. It is essential to note that such trenches pose a significant risk of collapsing during construction, and necessary safety measures must be taken during the construction work.

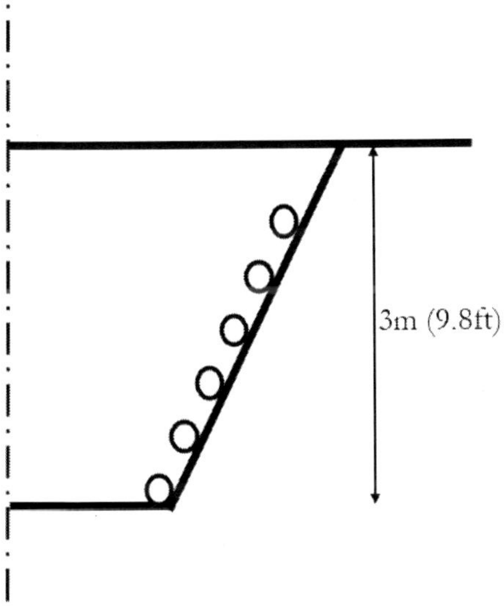

Figure 24: Example of a trench collector

4.2.3.6 Energy Fences

Energy fences are another option for obtaining heat (see Figure 25), being considered a special solution in the realm of geothermal energy sources. The energy fence is made of UV-resistant plastic pipes, partially buried in the ground, with the other part protruding above the ground. This allows the heat to be extracted from the outside air above ground and, if present, from solar radiation. The other part extracts heat from the ground. Energy fences are more suitable for milder climate zones but yield significant results in these conditions.

For covering the heating capacity during very low external temperatures, an electric auxiliary heater should also be installed. Another advantage of the special solution of the energy fence is its simultaneous use for property fencing or delineation, granting the energy fence a practical dual function in this regard. Admittedly, its aesthetic appeal might be up for debate.

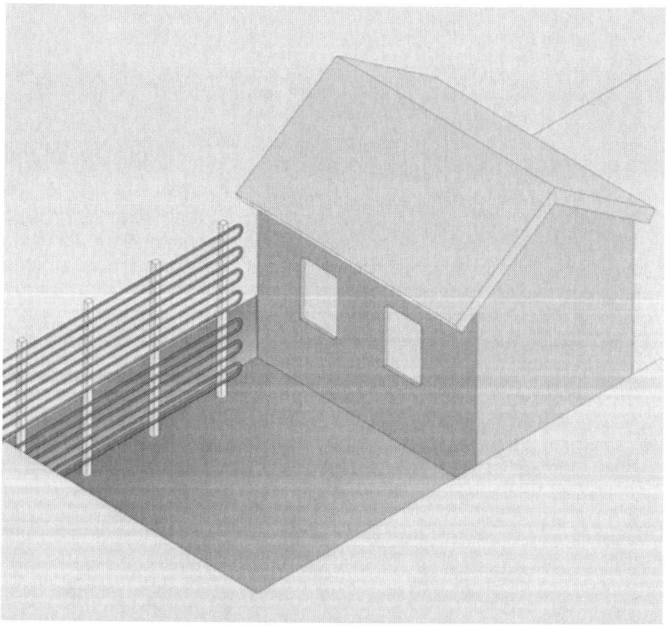

Figure 25: Example of an energy fence

4.2.3.7 Ground-Contact Concrete Elements

Another special solution in the realm of geothermal energy sources are ground-contact concrete elements used as heat exchangers. In this method, energy is extracted from the ground through large concrete components that are anchored vertically or horizontally in the ground. If energy is to be extracted from the ground through the foundation slab, attention should be given to maintaining a balanced heat budget or ensuring thermal regeneration. Ground-contact concrete elements are particularly suitable for preheating ventilation air in winter and pre-cooling it in summer.

4.2.4 Heat Source: Ground: Brine/Water-Heat Pump

The brine/water heat pump is also known as ground/water heat pump or geothermal heat pump. They derive heat from the ground and transfer it through the heated water circuit to the building via radiators or underfloor heating systems. Today, these heat pumps are mostly installed indoors, whereas in the past they were sometimes placed outdoors. Generally, these heat pumps are very commonly used and are particularly effective for building cooling (see Chapter 3.3).

4.3 Heat Source: Air

Especially when one considers retrofitting a heat pump, air as a heat source is often the first choice. Typically, outdoor air is used, which is always readily available and doesn't require a permit.

In general, it's advisable to install at least the evaporator of the heat pump outside the house. It's also possible to position the entire heat pump as a compact unit outdoors. It's important to note that when using air as a heat source, the airflow can produce noise. Therefore, always check if there are any special requirements according to building permits or permissions for installing an air heat pump.

The operation of the air heat pump can be explained similarly to Chapter 2. The heat source, air, provides an airflow that is channeled through the evaporator. Therefore, heat energy in the form of air is available. The still liquid heat carrier medium is warmed by this heat energy until it boils, then it condenses and is finally present as a gas. This is where the compressor comes into play. With the compressor, the heat carrier medium is compressed, and this increases the temperature of the gas. The heat exchanger then liquefies the heat carrier medium, extracting the heat energy from the medium. The extracted energy can be used for heating. A throttling process reduces the pressure of the now-liquid heat carrier medium, resulting in an expansion of its volume. With the volume expansion, the heat carrier medium cools down, and a temperature below the external air temperature is achieved. The heat pump process is completed and starts again.

In the heat source air, besides oxygen and nitrogen, water vapor is also present. The higher the air temperature, the more water vapor can be held in the air. Thus, at low temperatures, hardly any water vapor can be absorbed by the air. The air will then release the water vapor in the form of water, a process known as condensation, which occurs at the so-called dew point temperature. And this is where the problem arises. During heat transfer, the volume of air cools down as it passes through the evaporator, making it possible for the air to accumulate directly at the evaporator. This is further indicated by the accumulation of water droplets on the surface of the heat exchanger. Below 32°F, this moisture leads to ice forming on the heat exchanger. The icing adversely affects the heat transfer properties and, consequently, the entire heat pump cycle. This can even lead to the heat carrier medium not evaporating at all, which would result in a complete breakdown of the entire heat pump circuit.

This ice formation is counteracted with a defrosting function built into the heat pump, in the form of an electric auxiliary heater. This electric auxiliary heater requires additional energy, which, from an energy-saving perspective, is seen as a drawback. Nowadays, we have the capability to counteract the icing of the heat exchanger down to temperatures of 5°F using the electric auxiliary heater. However, the losses in the heat pump at such temperatures would be so high that it is recommended to use the air heat source only down to 23°F.

The higher the temperature of the air flowing into the evaporator, the better the performance of the air heat source.

In general, the air flowing to the evaporator should be above 32°F. This clarifies why it's advisable to use heat pumps with the air heat source in a bivalent operation mode (the heat pump operates in conjunction with a second heat generator (refer to Chapter 5.3 for more details)).

So-called energy tubes, which are laid about 2m deep in the ground, are designed to warm the incoming air to around 50°F, the constant temperature in the ground. The air flowing to the evaporator is only drawn through the energy tube when the temperature of the surrounding ground is higher than the air temperature. Otherwise, the heat pump should draw in directly, without the energy tube. This process can be automatically controlled.

In the next step, we want to specifically address the various types of heat pumps from the air heat source.

4.3.1 Heat Source: Air: Heat Pumps

4.3.1.1 Compact Design Air-to-Water Heat Pump

A **compact design air-to-water heat pump** can be installed both indoors and outdoors. Heat pumps in compact designs are also referred to as **monoblock designs**.

In the compact design, the condenser is directly integrated into the device. The entire heat pump is installed outdoors in the case of an outside installation (see Figure 26 and Figure 27), with only the hot water tank and hydraulic station located inside the building.

The heating water can be led into the building via above-ground or underground connecting pipes. The transfer of thermal energy from the condenser to the storage for supply and return is implemented using the heat carrier water. The advantage of this type of heat pump is that they can be directly installed by the heating installer (system mechanic for sanitary, heating, and air conditioning technology) and there's no need to specifically involve a refrigeration technician (for the specialties, see Figure 51). A refrigeration technical qualification is also not necessary. However, when installing an air-to-water heat pump outside, it's important to note that the heating water inside the heat pump requires antifreeze protection. Freezing is usually prevented by the integrated auxiliary heating, which in turn requires additional energy. If, however, there's a prolonged power outage during extreme cold, the condenser might freeze, damaging the heat pump. Especially given today's increasing risk of blackouts, this aspect shouldn't be overlooked. Typically, the heat pump isn't covered by household insurance. To completely eliminate the risk of freezing, the circuits are decoupled via an intermediate heat exchanger. As a result, an approved antifreeze is added to the liquid circuit located between the heat pump and the intermediate heat exchanger in the building. However, this safety device does result in energetic losses in the heat pump.

Figure 26: Monoblock-design, Eureka-Luft (EU-L) - series by LAMBDA Wärmepumpen GmbH

When retrofitting an air-to-water heat pump in a compact design, it's important to note that, if installed outside, it must be placed on a designated foundation. Furthermore, a frost-free condensate drain is required, and the heating water connection line from the heat pump to the house must be insulated.

If an indoor installation of the air-to-water heat pump is desired, relatively large openings in the outer wall are required for the pump's intake and exhaust, depending on the heating power (e.g., 900 x 900 mm). This aspect is particularly important to consider from a structural standpoint.

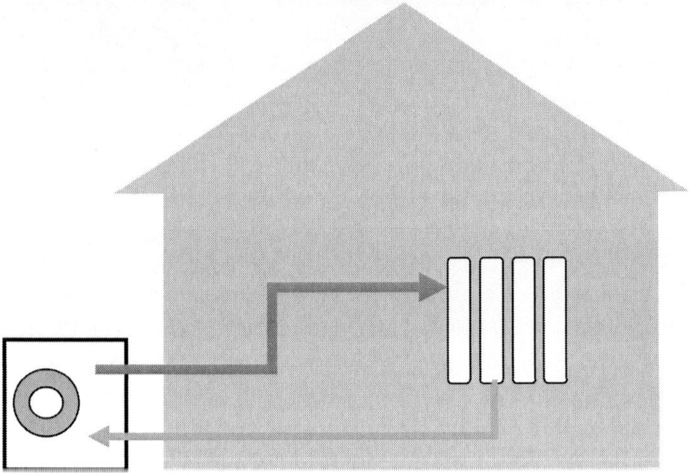

Figure 27: Operating principle of monoblock design, outdoor installation

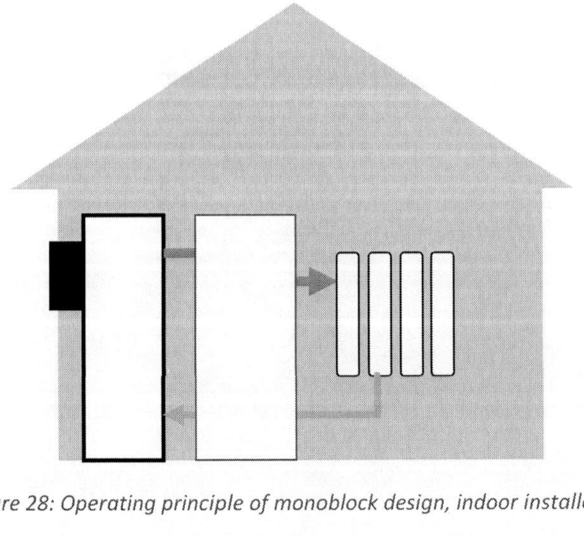

Figure 28: Operating principle of monoblock design, indoor installation

4.3.1.2 Air-to-Water Heat Pump in Split Configuration

Such a heat pump is an air-to-water type where one part is installed outside and the other inside the building. Because of this separation into an outdoor and an indoor unit, it's referred to as a **split configuration** (see Figure 29). Noise-generating components like the fan, heat exchanger (evaporator), and compressor can be placed outside, while the indoor unit with the condenser, circulating pump, and control technology can be installed inside the house due to its compact size. In some designs, the compressor is also located inside the building. Through the outdoor unit, external air is drawn in, which causes a refrigerant circulating within the heat pump to evaporate based on its temperature. The temperature of the evaporated refrigerant is further increased by compression using the compressor in the outdoor unit and then directed through lines to the indoor unit. This also ensures that no water is present outside, eliminating the need for antifreeze. In the indoor unit, the thermal energy of the refrigerant is transferred to the heating system at the condenser, causing the refrigerant to condense again. Once the liquid refrigerant is directed back outside, the cycle begins anew. Unlike the air-to-water heat pump in a compact configuration, for the air split heat pump, a heating engineer or refrigeration technician with the appropriate refrigeration qualification is required (for expertise areas, see Figure 51). The air split heat pump is currently gaining significant traction in new builds.

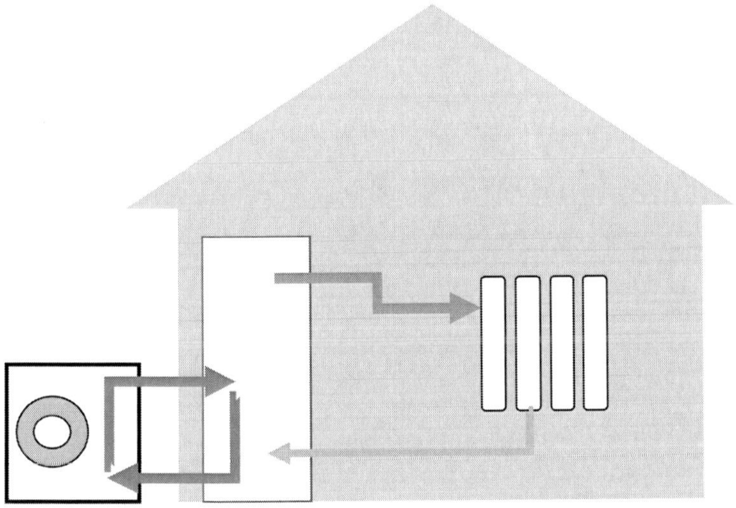

Figure 29: Operating principle of split design

4.3.1.3 Air-to-Water Head Pump with Split Brine Heat Exchanger

In contrast to the air split heat pump, the heat exchanger outside the building in the **air-to-water heat pump with a split brine heat exchanger** is filled with brine (see Figure 30). This brine is warmed by the outside air and, at a laying depth of more than 80 cm, is directed inside PE pipes to the evaporator of the heat pump in the building.

The underground PE pipes have the added benefit of initiating warming of the air inside, similar to an earth collector.

One advantage of this type of heat pump concerning noise reduction is that the air/brine collector can be set up to 30 meters away from the heat pump installed in the building. The fan speed can be reduced overnight, leading to further noise reduction. There's typically no risk of freezing, even in the event of a power outage, thus ceasing the operation of the electrical heating on the external unit.

Since no refrigerant is used, an air-to-water heat pump with a split brine heat exchanger can be installed by a heating engineer without refrigeration qualification and without needing a refrigeration technician (for expertise areas, see Figure 30).

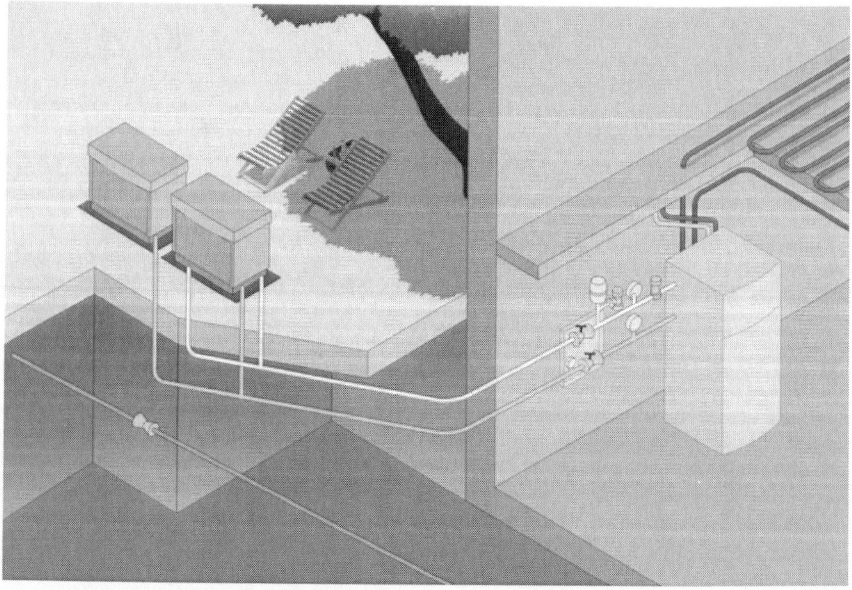

Figure 30: Operating principle of split construction

4.3.1.4 Domestic Hot Water Heat Pump

A domestic hot water heat pump is essentially an air-to-water heat pump that doesn't draw its heat from the external ambient air, but from warm, used indoor air (circulated air) or exhaust air. The domestic hot water heat pump differentiates between circulated air/domestic hot water heat pump and exhaust air/domestic hot water heat pump (see Figure 31). Inside the heat pump, there is an integrated fan which directly draws room air or through a piping system. Typically, this is a small heat pump mounted on and integrated with the hot water storage tank that extracts heat from the surroundings and uses it to heat water. This is referred to as heat recovery. For instance, one can effectively use the warm exhaust air from an industrial ventilation system as supply air for the evaporator. Additionally, supply air from heating rooms, conservatories, kitchens, bathrooms, or basements can also be used as intake for the evaporator. Notably, a positive aspect is that for domestic hot water heat pumps that draw supply air from the basement, simultaneous dehumidification of the basement occurs. Moreover, if necessary, such a heat pump includes an electric heating element or possibly even a connection to a solar system. This type of heat pump is generally very suitable for combination with photovoltaic systems, as the self-generated electricity is sufficient for the comparatively low power of the domestic hot water heat pump.

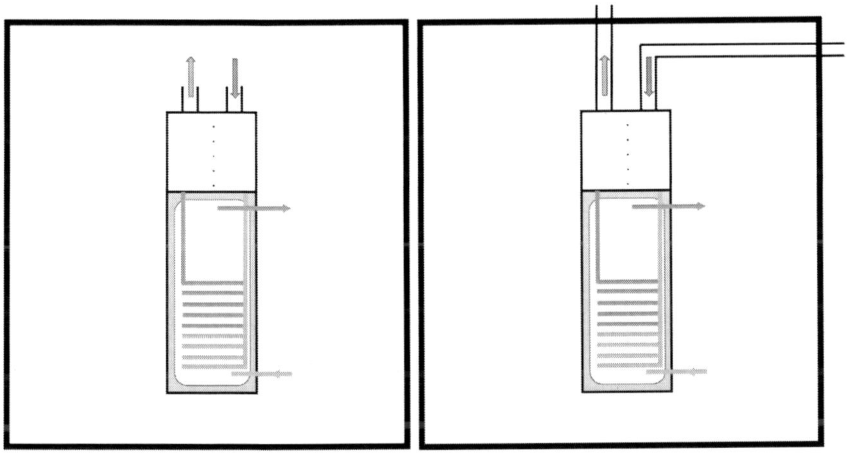

Figure 31: Air circulation/hot water heat pump (left) and exhaust air/hot water heat pump (right)

4.3.1.5 Air-to-Air Heat Pump

The **air-to-air heat pump** does not require a conventional heating surface (e.g., in the form of a radiator) within the building but heats the interior spaces exclusively through the air. Simply put, it's akin to an air conditioning unit with a heating function. By heating the space through warm air (convection), a rapid warming of the premises occurs. This straightforward and space-saving installation makes it particularly appealing for retroactive installations and offers lower investment costs. However, the noise generated by the fan can be bothersome. Similarly, the absence of heat radiation (thermal radiation) is often perceived as less pleasant, and the warmth from this type of heat pump tends to dissipate relatively quickly after the air-to-air heat pump is turned off. This type of heat pump is commonly used in hotel rooms.

4.4 Additional Designs

In addition to the special designs of energy absorbers already mentioned in the section on ground heat sources, there are other designs for both the heat sources and the heat pumps themselves that should not go unmentioned here.

4.4.1 Additional Heat Pump Designs

4.4.1.1 Hybrid Heat Pumps

Hybrid heat pumps combine different heat sources. This means that the pump is, for example, a combination of a gas condensing boiler and an air or ground source heat pump.

An advantage here is that depending on the current price level of the energy source, it can be preferentially used. The hybrid heat pump operates similarly to a bivalent mode (see Chapter 5.3), which allows for heating circuits with different system temperatures.

4.4.1.2 Heat Pumps with multiple Heat Sources

With this type of heat pump, it is possible to combine different heat sources within a single heat pump.

Currently, a heat pump model called "System Zero" developed by the company Dimplex from Kulmbach (Germany) is available on the market. This system allows the use of both air/water and brine/water.

With this, the heat sources air and brine can also be combined simultaneously, if this proves to be the most fruitful at any given time. The heat pump autonomously regulates this. This is feasible because the system integrates two circuits, namely two evaporators. In addition to heating, this heat pump also offers the capability to cool.

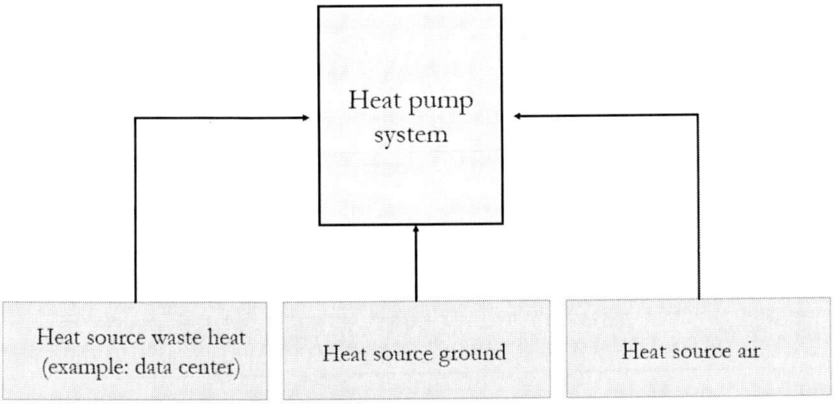

Figure 32: Functioning principle of a heat pump with multiple heat sources

4.4.1.3 Heat pumps with direct Evaporation and direct Condensation

Where, for instance, with the ground heat source, the absorbed heat is typically transferred to a brine (frost-protected heat carrier fluid) from the ground and subsequently passed on to the refrigerant in another heat transfer process, this is not the case with a heat pump featuring direct evaporation. With the heat pump that uses direct evaporation and direct condensation, the process is executed within one loop. Using the ground example, the refrigerant from the refrigeration cycle is directly passed through the ground probe and the underfloor heating system. The loss from the heat exchanger of a few degrees Fahrenheit is thereby avoided, and as a result, the heat pump with direct evaporation achieves a better (Heating) Seasonal Performance Factor (SPF / HSPF).

The downside to this is the complicated regulation of heat in individual rooms. Furthermore, direct evaporation and direct condensation require a larger amount of refrigerant, leading to increased costs and a heightened probability of leaks.

4.4.1.4 High-Temperature/ Very High Temperature Heat Pumps

High-temperature heat pumps (HTHP) refer to heat pumps with a sink temperature (flow temperature) ranging between 176 and 212°F.

If the flow temperature exceeds 212°F, it is referred to as very high-temperature heat pumps (VHTHP). The precise classifications based on sink and source temperatures are depicted in Figure 33.

The fundamental distinction between HTHP and VHTHP in comparison to a conventional heat pump (HP) lies in their operational limits. Today, high-temperature heat pumps are primarily utilized in the food processing, paper, and chemical industries for processes such as drying, sterilization, evaporation, and heat recovery.

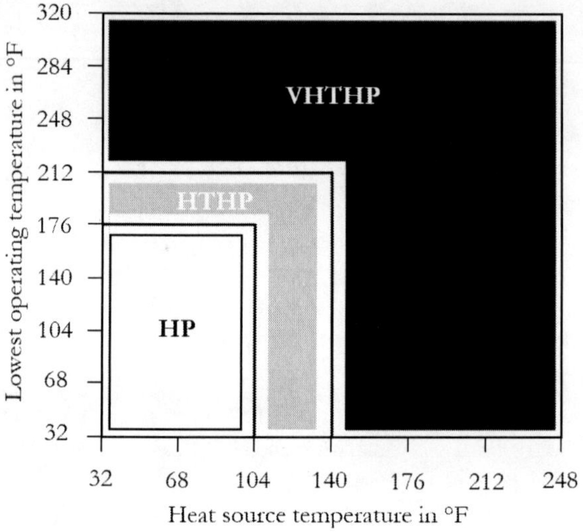

Figure 33: Classification of heat pumps according to temperature

They often deliver large heating capacities ranging from 20kW to 20MW. Their coefficient of performance ranges between 2.4 to 5.8 with a temperature differential of 40 to 95 degrees Fahrenheit. In the past, smaller high-temperature heat pumps with a heating capacity of up to approx. 90kW were predominantly used in existing buildings to avoid the need to replace the existing radiators.

Research on high-temperature heat pumps is ongoing and continues for the existing building sector. Particularly for existing buildings, high-temperature heat pumps are of interest since it won't always be feasible to retrofit radiators to surface heating systems, which would only require a lower sink temperature.

4.4.1.5 Pool-Heat Pumps

When using a pool heat pump (swimming pool heat pump) in the summer to heat one's pool, depending on the air and water temperature, the heat pump can deliver up to five times the electrical power it consumes in the form of heat. The pool heat pump operates on the same principle as conventional heat pumps, extracting energy from the surroundings and delivering it in a concentrated form to the pool water. In this process, the pool water circulates in the secondary loop. The pump doesn't run continuously but only activates when the pool's temperature drops below a certain level. When combined with a PV (photovoltaic) system, the power requirements of the pool heat pump can be met by the electricity available during the summertime (more on the combination in Chapter 10).

4.4.2 Additional Heat Source Designs

4.4.2.1 Heat Source: Wastewater

As the name suggests, the necessary heat energy is extracted from wastewater and transferred into heat by the heat pump (see Figure 34). This type of heat source is primarily used for residential units, settlements, or districts via heating networks. If there is a relatively short distance between the heat source and the point of consumption, the heating energy is raised to the necessary temperature level by a large heat pump and then transported a short distance to the point of use. This is known as district heating. If there are greater distances between the heat source and the point of consumption, the heat is first transported cost-effectively at the output temperature level, and then processed directly at the point of consumption. This is then referred to as cold district heating.

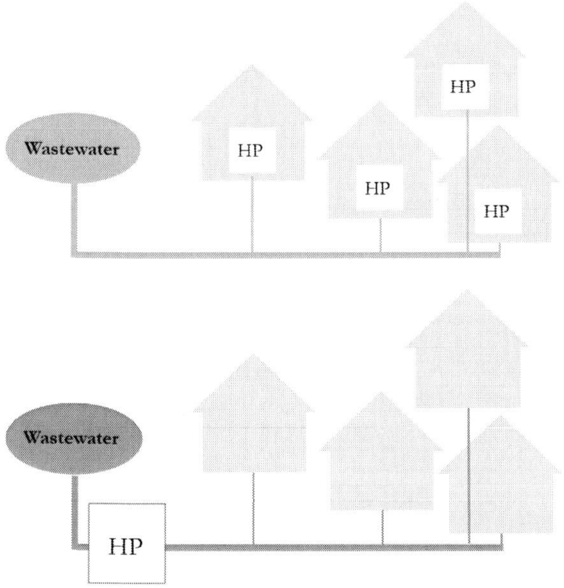

Figure 34: Illustration of district heating and cold district heating

4.4.2.2 Ice Storage

This represents a relatively new development that has proven to be highly efficient.

The system consists of three components: a water tank (the ice storage), the brine/water heat pump, and a solar air absorber.

But first, the question arises: what exactly is an ice storage? An ice storage is a latent heat storage. The term "storage" in this context should not be confused with other storage options, such as a buffer tank. In the case of the ice storage, a large concrete container is positioned underground and filled with water. This can be implemented in the garden of a single-family home. The ice storage utilizes the latent heat released during the phase transition from water to ice. As water transitions into ice, crystallization energy is generated, releasing significant amounts of energy.

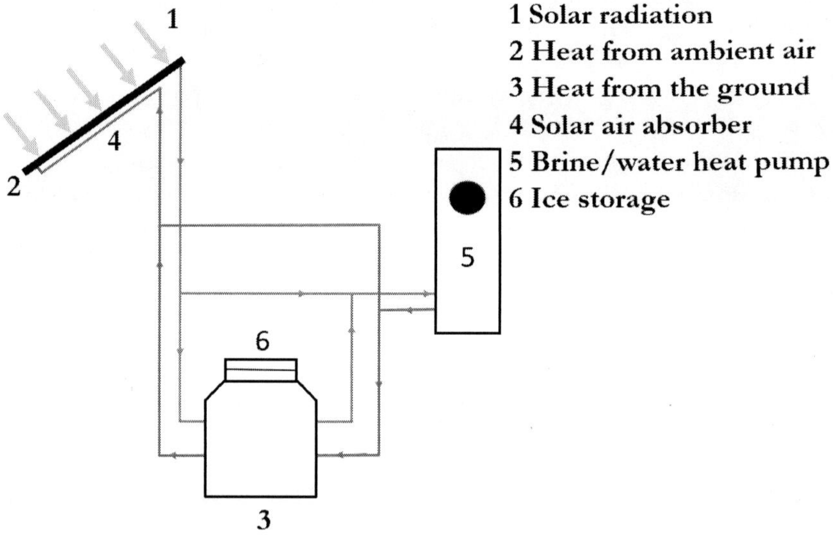

Figure 35: Illustration of the Concept of an Ice Storage System

When it becomes cold outside, the heat pump initially uses the energy from the solar air absorber for heating. If the external temperature drops below 32°F, this energy becomes insufficient, and the energy from the ice storage is harnessed with the heat pump until the ice storage is completely frozen. The extracted heat is then used to warm up the heating circuit and, if desired, the hot water circuit.

The frozen water in the ice storage is gradually thawed by the geothermal heat through the concrete shell. To accelerate this, a solar air absorber installed on the roof, which operates both with solar heat and ambient air heat, is used. It's worth noting that the concrete container in the ground is not insulated, so the geothermal heat also contributes to the thawing process. Through the solar air absorber, the water in the ice storage can be heated so significantly by the sun's heat and ambient air that it results in very high performance figures.

During the summer, the thawing process can be halted, and the frozen ice storage can be used to cool the house by releasing energy through the home's heating surfaces. Figure 35 illustrates the concept.

Summary:

There are three **fundamental sources of heat: water, ground (brine)**, and **air**. Water is the most efficient source of heat, while the air source benefits from its unlimited availability. With the ground source, there are fundamentally two different methods to harness geothermal energy. The first method uses geothermal probes (ground probes) that delve deep into the earth to draw on the Earth's heat. The second method involves ground collectors, which, instead of going deep, are laid out across the surface underground. However, there are also other design forms for energy sourcing, such as trench collectors, geothermal baskets, or ditch collectors.

5 OPERATION MODES OF HEAT PUMPS

The different operation modes of heat pumps are crucial for their function and efficiency, and this chapter aims to provide you with more insights about them.

There are three primary operation modes for heat pumps. These are the **monovalent mode**, the **monoenergetic mode**, and the **bivalent mode**. Once the operating mode has been determined, there are two more selection criteria to consider: the design point and the degree of coverage. The **design point** indicates the lowest outdoor temperature at which the heat pump can provide the required heating energy. The degree of coverage specifies what percentage of the annual required heating energy can be supplied by the heat pump. Once the design point and the degree of coverage have been evaluated, the preferred type of heat source and the heat pump system can be determined. When choosing the heat source, local conditions concerning space, soil properties, and the maximum possible costs are also crucial factors. Thus, the operation mode, the heat source, and the chosen heat pump system need to be determined in an iterative process until the optimum is found. This chapter will explain each operating mode in detail.

5.1 Monovalent Operation Mode

In the monovalent operation mode, the entire building supply is managed by the heat pump.

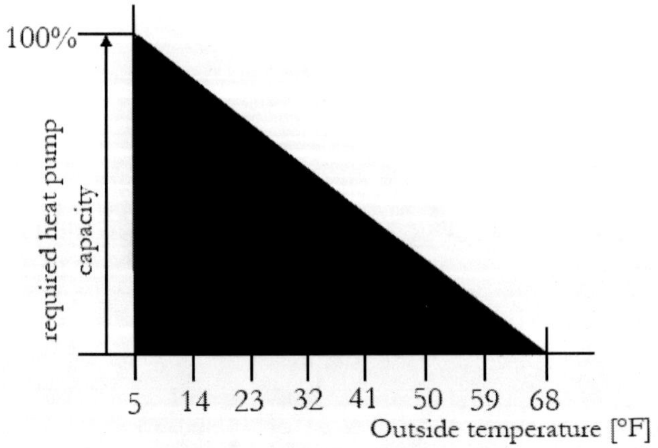

Figure 36: Representation of Monovalent Operating Mode

A heating boiler thus becomes obsolete. With the monovalent operation mode, a coverage rate of 100% is to be achieved, with the design point being set to the locally specified lowest outside temperature. Figure 36 illustrates the monovalent operation mode.

For this mode of operation, the building must be energetically very well tuned. This is typically more suitable for new buildings or low-energy buildings. To provide the required energy all year round, the heat sources water or earth should be used (see Chapter 4).

5.2 Monoenergetic Mode of Operation

The monoenergetic mode of operation takes over the primary heat supply in the building. However, unlike the monovalent operation mode, it does not cover the peaks during very cold outside temperatures. Such peaks are then taken over by an auxiliary heater operating on the same energy basis (e.g., an electric heating element). Thus, the design point is shifted to the right to a lower slope of the curve, aiming to achieve a coverage rate of about 95%. In general, the system is designed in such a way that the auxiliary heater has to cover between 5 to 15% of the annual heating demand. The point at which the auxiliary heater is called upon is the design point. This coverage rate is still sufficient for most temperatures, with occasional dips into lower temperatures covered by adding the auxiliary heating. However, the monoenergetic operation mode also requires good energy coordination in the building (e.g., insulation), making it more suitable for new buildings or low-energy constructions.

5.3 Bivalent Mode of Operation

A bivalent mode of operation includes, in addition to the heat pump, another heat generator that is activated to heat the building when temperatures drop below the design point. This additional heat generator can be a conventional oil or gas heating system. In the bivalent mode of operation, the design point is referred to as the bivalence point. Within the bivalent mode of operation, there are three categories: the bivalent parallel mode of operation, the bivalent alternative mode of operation, and the bivalent partially parallel mode of operation.

5.3.1 Bivalent Parallel Mode of Operation

The first category, the bivalent parallel mode of operation, differs from the classical bivalent mode in that, upon reaching the design point, the secondary heat generator operates in parallel with the heat pump to jointly cover the heating demand (and not just to support it (see Figure 37).

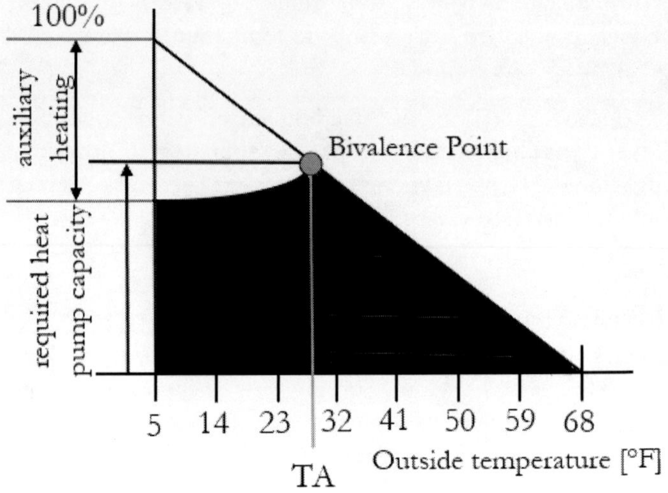

Figure 37: Representation of Bivalent Parallel Operating Mode; TA = Activation Temperature for Supplementary Heating

5.3.2 Bivalent Alternativ Mode of Operation

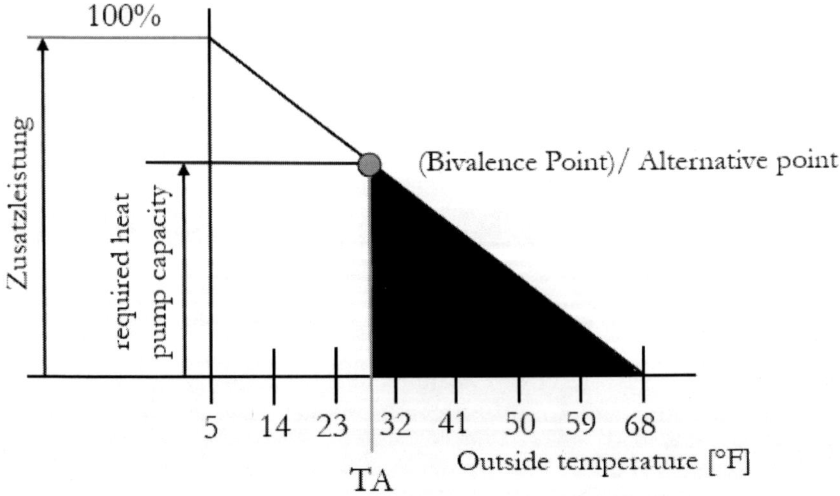

Figure 38: Representation of Bivalent Alternative Operating Mode; TA = Activation Temperature for Second Heat Generator and Deactivation of First Heat Generator

The bivalent-parallel mode of operation is frequently used in renovation measures or in older buildings, making use of the already installed secondary heat generator, such as an oil or gas boiler, for this purpose.

Another category of the bivalent mode of operation is the bivalent alternative mode (see Figure 38). Here, upon reaching the design point, the heat pump shuts off completely, and the secondary heat source operates alone to provide the entire heat supply.

This mode of operation is also more commonly seen in renovation projects or older buildings. However, in comparison to the bivalent-parallel mode, even more of the heat is provided by the secondary heat generator, which ideally shouldn't be the aim today.

For large systems, this mode is often used for hot water preparation.

5.3.3 Bivalent Partial Parallel Mode of Operation

A third category of the bivalent mode of operation is the bivalent partial-parallel mode (see Figure 39). In this mode, the heat pump alone covers the heating load of the property up to the bivalence point. Below the bivalence point, the secondary heat generator is additionally activated. Thereafter, the heat pump operates in parallel with the secondary heat generator until the maximum flow temperature is reached, at which point it is then turned off. As with the bivalent alternative mode of operation, the secondary heat generator is designed to handle the entire heating load.

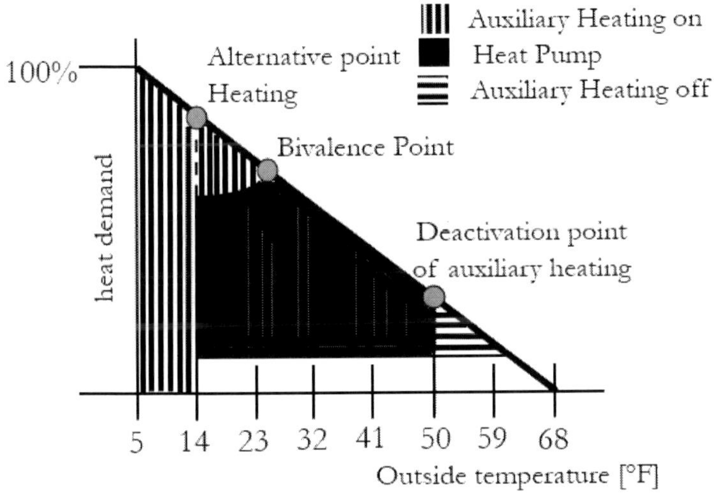

Figure 39: Representation of Bivalent Partial Parallel Operating Mode

Summary:

For heat pumps, there are three distinct (primary) modes of operation. These are the **monovalent mode**, the **monoenergetic mode**, and the **bivalent mode**.

6 BUFFER AND HOT WATER STORAGE

In this chapter, we will first explain the definition of buffer and hot water storage, the function, operation modes, and types of buffer storage, the function of the hot water storage, and finally, the combined storage.

Since there is often confusion or misunderstandings about these terms, we want to differentiate them precisely first. The term "hot water storage" is the umbrella term for storing warm water. The hot water storage can store heating warmth as well as potable water warmth or even additionally heat the potable water. When the hot water storage is integrated into the heating system, storing heating water, it is referred to as a buffer storage. Thus, a hot water storage can be defined as storage for the heating water (explicitly buffer storage) and as storage for potable water (where only the term hot water storage is used).

I tried to make the translation as clear and accurate as possible. Please review to ensure it aligns with the intended meaning and technical specifications of your original text.

6.1 Buffer Storage (Heating Water)

Buffer storages can perform three functions:

- Overcoming energy supplier's downtime
- Hydraulic protection (decoupling or provision of minimum volume)
- Optimization of the heat pump system runtime

When the buffer storage is designed to **bridge downtime**, the volume of the selected buffer storage always suffices for the required provision of the minimum volume in the distribution line or for runtime optimization. Therefore, designing the buffer storage for downtime bridging can be seen as encompassing all three possible functions. To ensure a smooth operation of heat pumps, a **minimum flow of heating water** must always be guaranteed. For this reason, the use of a buffer storage must be checked. The buffer storage can be used for the hydraulic decoupling of the flow rates from the heat pump circuit and the heating circuit. This becomes necessary when the flow in the heating circuit is reduced by thermostats, and the heat pump circuit should remain constant at the same time. To maintain the efficiency and lifespan (compressor lifespan) of the heat pumps, frequent switching on and off (cycling) of the heat pump should be avoided. This frequent cycling is prevented by adequately sized buffer storage (**runtime optimization**).

For air-to-water heat pumps, the buffer volume is also used for the defrosting process of the evaporator. Depending on the legal regulations, there can be **downtimes** during peak times of electrical energy supply. But what does the energy suppliers' downtime mean? At times of high electricity consumption with energy suppliers, known as peak loads, it becomes uneconomical for the energy suppliers to continue providing cheap electricity. Therefore, during these times, the discounted heat pump tariff faces a predetermined downtime and is thus disconnected from the grid. The heat pump's buffer volume can be used to bridge these downtimes. When combining the heat pump with a photovoltaic system, the heat pump can be powered with self-generated electricity, and the excess energy can be stored as hot water in the buffer storage. The buffer storage can also balance peak loads. Detailed calculations on the various uses are listed in Chapter 7.7. For a wet-installed underfloor heating (see also Chapter 7.3), a buffer storage is not necessarily required, as the screed acts as a buffer in this case. Buffer storage applications are divided into **serial** or **parallel buffer storage**.

Serial buffer tanks (Figure 40 and Figure 41) are recommended for systems with only one heating circuit and serve to increase the volume of the heating system. Compared to parallel buffer tanks, they are energetically more favorable but more demanding in terms of hydraulics. When installed in series to ensure the minimum flow rate when all heating circuits are closed, a bypass valve must be added. It is essential to distinguish whether the buffer tank is installed in the feed line or the return line.

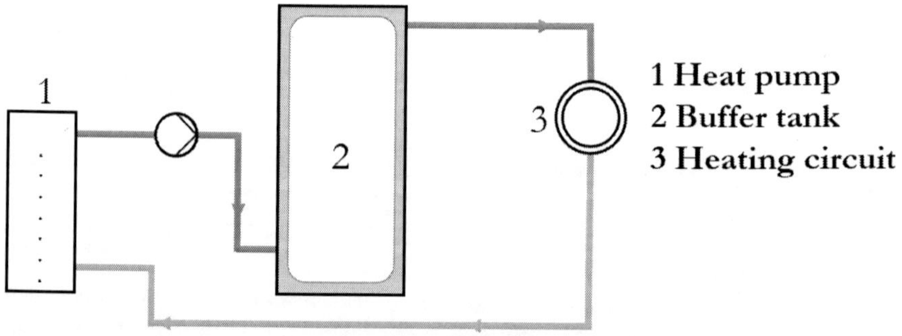

Figure 40: Series Buffer Tank in the Flow

When the buffer tank is integrated into the feed line, there is the opportunity to incorporate a heating element as an additional heat generator into the buffer tank. When installing the buffer tank in the return line, due to the lower temperature of the return flow compared to the feed flow, lower radiation losses from the buffer tank can be expected. This type of installation of the buffer tank in the return line is predominantly used to extend the compressor's runtime.

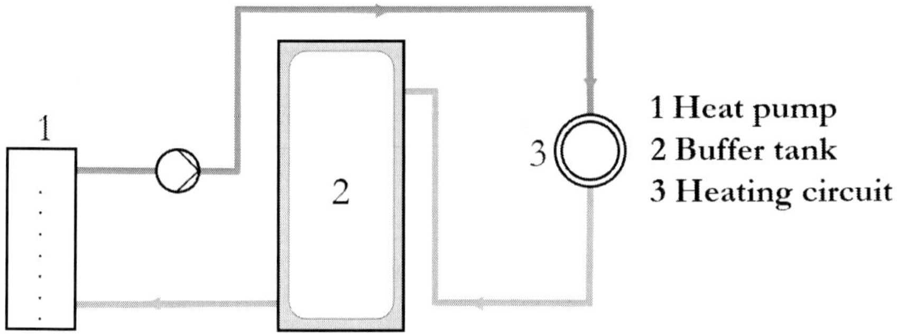

Figure 41: Series Buffer Tank in the Return

For **parallel buffer tanks** (Figure 42), the heat pump is hydraulically decoupled from the heating circuit, fulfilling the function of a hydraulic interface. With the parallel buffer tank and the accompanying hydraulic decoupling, multiple heating circuits can be used in parallel. Through parallel buffering, hydraulic errors are most reliably avoided, and the required minimum volume of the heat pump is always ensured, without being dependent on the flow rate of the heating circuit. However, with parallel buffer tanks, an additional circulation pump with energy demand is necessary in the producer circuit.

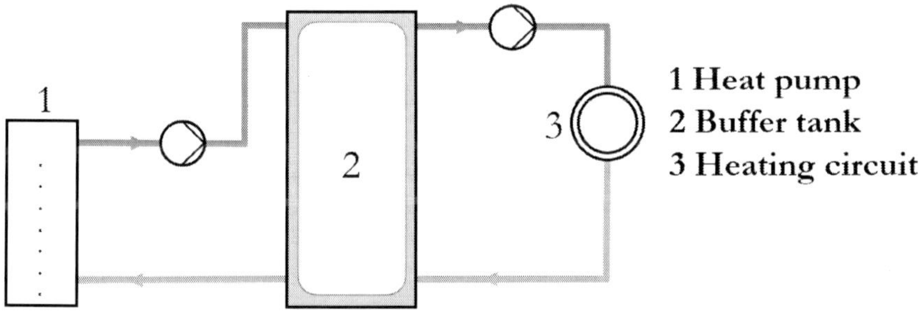

Figure 42: Parallel Buffer Tank

Another downside is that, on average, the heat pump in parallel buffer tanks operates at slightly higher feed temperatures, as the heating circuit with the highest temperature level determines the target temperature in the buffer tank.

6.2 Drinking Water Storage and Drinking Water Heating (Domestic Hot Water Preparation)

6.2.1 Fundamentals of Drinking Water Storage and Drinking Water Heating (Domestic Hot Water Preparation)

In the following, various **types of domestic hot water storage** for heating and drinking water heating with heat pumps will be introduced. However, an explanation of the basic differences is required beforehand. There are **open and closed domestic hot water storage systems.**

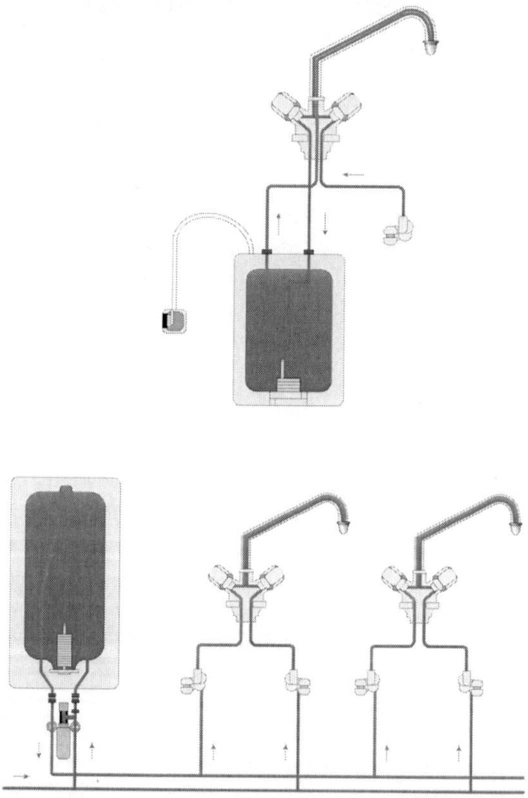

Figure 43: Open Hot Water Tank (Top) and Closed Hot Water Tank (Bottom) Principle

With an open domestic hot water storage, there is no water line pressure, and it is a supply near the point of consumption. Only one tap can be supplied. They are usually small, located directly next to the tap, and can be found, for example, at sinks (Figure 43). They are generally used only for drinking water storage. Closed domestic hot water storage systems can supply multiple taps. The closed domestic hot water storage is subject to the water line pressure.

The domestic hot water storage units relevant to the storage of heating water as well as drinking water storage and heating with heat pumps are closed systems (Figure 43).

For domestic hot water preparation (drinking water), a distinction is made between **decentralized** and **centralized domestic hot water preparation**. In decentralized domestic hot water preparation, the water heater is usually installed directly at a withdrawal point (bathtub, sink, etc.). Centralized domestic hot water preparation is usually implemented in single or multi-family houses as a centralized domestic hot water preparation system and takes place via a common piping network from a central location (boiler room).

Furthermore, in terms of domestic hot water storage, a distinction is made based on the method of water heating: direct heating (directly heated domestic hot water storage) and indirect heating (indirectly heated domestic hot water storage). With directly heated domestic hot water storage, the storage unit itself acts as an independent heat source, heating the water directly inside the storage tank. Often, the directly heated domestic hot water storage is also referred to as a boiler. Typically, such a system is used for decentralized domestic hot water preparation.

When using heat pumps, the system involves closed domestic hot water storage, centralized domestic hot water preparation, as well as indirect heating (indirectly heated domestic hot water storage).

6.2.2 Systems of Domestic Hot Water Storage and Heating (Domestic Hot Water Preparation)

The operating methods described in Chapter 5 are also applicable to domestic hot water heating or domestic hot water preparation systems. Regarding the design, there are the following five systems:

- Monovalent Domestic Hot Water Preparation System
- Monoenergetic Domestic Hot Water Preparation System
- Bivalent Domestic Hot Water Preparation System
- Storage Charging System
- Central Flow-Through System

6.2.2.1 Monovalent domestic hot water system

In a monovalent domestic hot water system, the heat pump, as the sole heat generator, covers the entire demand for domestic hot water.

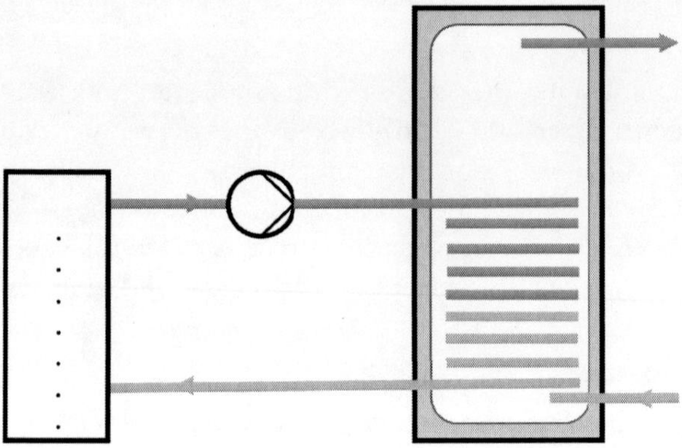

Figure 44: Monovalent Domestic Hot Water Preparation System

Here, the heat energy of the heating water is transferred to the hot water in the domestic water storage tank via a coil (see Figure 44). The simple system design can be seen as an advantage. A drawback is that such a domestic hot water system requires a high supply temperature from the heat pump of 158°F to ensure a drinking water temperature of 140°F.

6.2.2.2 Monoenergetic Domestic Hot Water System

In the monoenergetic hot water preparation, the heat pump covers the entire domestic hot water demand until it reaches the maximum flow temperature.

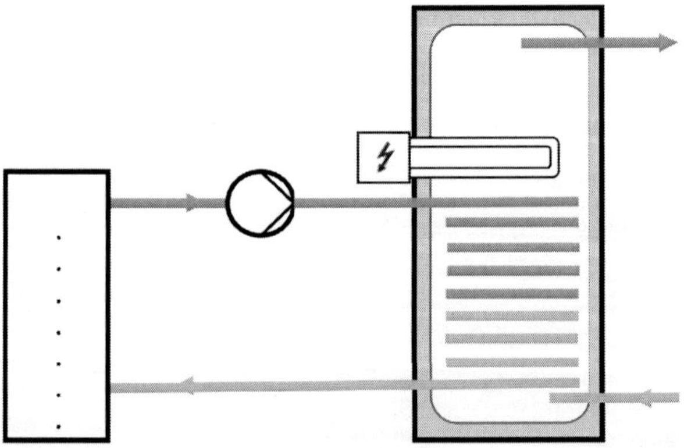

Figure 45: Monoenergetic Domestic Hot Water Preparation System

Any additional heating requirement is ensured electrically by heating elements (see Figure 45). Here too, the heat energy from the heating water is transferred to the hot water in the domestic water tank via a coil. While this solution is cost-effective, it can only achieve the required or desired domestic water temperature of 140°F using heating elements.

6.2.2.3 Bivalent Domestic Hot Water System

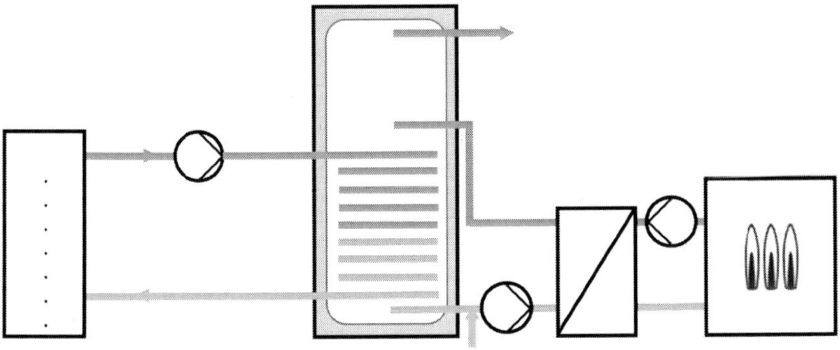

Figure 46: Bivalent Domestic Hot Water Preparation System

In the bivalent domestic hot water system, the heat pump covers the base load. The range that is uneconomical for the heat pump is covered by another heat generator (e.g., oil heating) (see Figure 46). Here too, the heat energy from the heating water is transferred to the domestic hot water in the storage tank through a coil. This solution offers very efficient operation while maintaining a high hot water performance (140°F). However, the high investment and operating costs can be seen as a disadvantage.

6.2.2.4 Storage Charging System

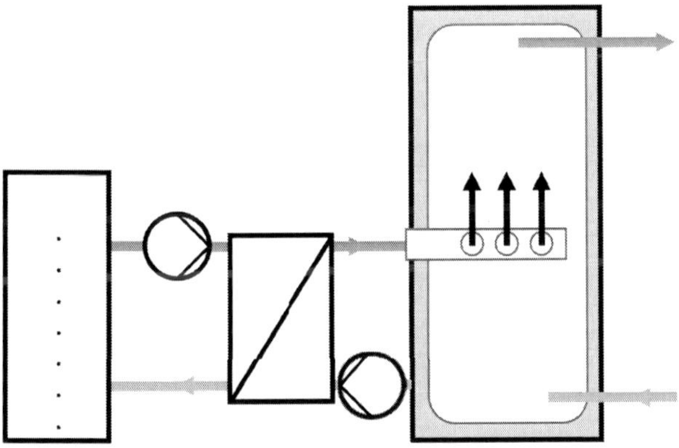

Figure 47: Storage Charging System

In the storage charging system, charging storage tanks with external heat exchangers (most commonly plate heat exchangers) are used.

Due to the external heat exchangers, the surface area of the heat exchanger can be chosen to be as large as desired, allowing for high performance levels.

These charging storage tanks are equipped with charging lances, which reduce the inflow speed of the heated water. Thus, even with the small temperature difference of heat pumps, the desired domestic hot water temperature can be achieved, and the heat pump covers the base load. The range that is uneconomical for the heat pump is covered by another heat generator (see Figure 47). Here too, the heat energy from the heating water is transferred to the domestic hot water in the storage tank through a coil. This solution offers very efficient operation while maintaining a high hot water performance (140°F). However, the high investment and operating costs can be seen as a disadvantage.

6.2.2.5 Central Flow-Through System

Another option is the central flow-through system. In the central flow-through system, the storage tank contains heating water instead of potable water. The drinking water is routed outside the tank through a heat exchanger that is permeated with heating water and is thereby heated.

6.2.3 Combined Storage Tank

Another option is the combined storage tank, which stores the **heating water (buffer tank) and potable water (hot water storage) in a single container**. However, there are special requirements for installing a combined storage tank. Its advantage, due to the combination, is its small footprint.

Here, two basic types of combined storage tanks are differentiated:

- Combined storage tank with integrated flow-through principle
- Tank-in-tank combined storage

Combined storage tanks with internal heat exchangers (also known as the integrated flow-through principle) are equipped with heat exchanger tubes within the container, which serve as a system separation between the circuits of the heating water and the potable water, as well as for the heating of the drinking water by the heating water. Due to the internal heat exchangers, this type of combined storage tank has an increased weight, which makes installation on-site more challenging. Figure 48 shows such a combined storage tank, where the drinking water heating is powered by the heat pump.

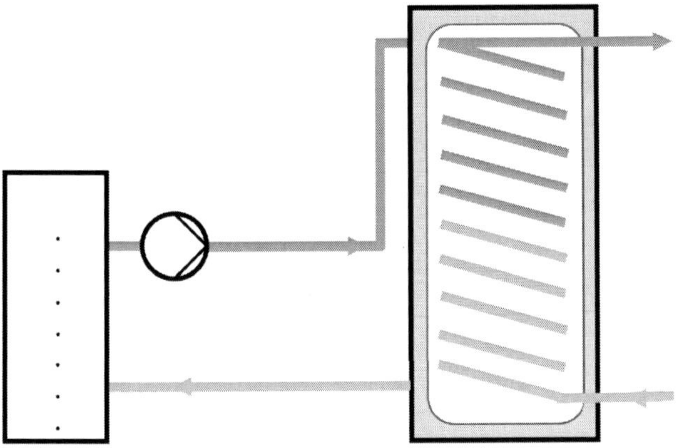

Figure 48: Combination tank with internal heat exchanger

The **combined storage tank with the tank-in-tank system** (see Figure 49) ensures heat transfer through the tank wall. Here, the potable water storage (the inner tank) is surrounded by the buffer tank containing the heating water (the outer tank). The potable water storage is generally designed to be smaller in size and is typically suitable only for a limited hot water demand.

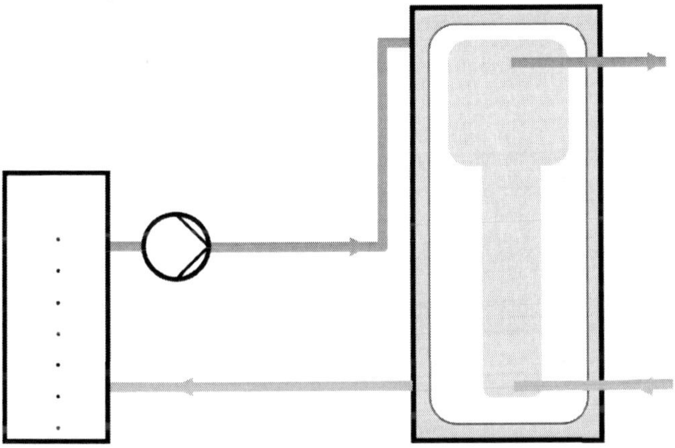

Figure 49: Combination tank with tank-in-tank system

The principle is similar to the double-shell storage tank. Figure 49 illustrates the system. The amount of energy produced by the heat pump is directed to the potable water heating based on the temperature level, either in the form of stratification or internal partitions of the tank.

Summary:

Buffer storage tanks have three functions: **bridging the lockout period** of the energy supplier, **hydraulic security** (decoupling or providing a minimum volume), and optimizing the **runtime of the heat pump system**. The operating modes described in Chapter 5 are also applicable to potable water heating and hot water preparation systems. In terms of design, the following systems are available: monovalent hot water preparation system, mono-energetic hot water preparation system, bivalent hot water preparation system, storage charging system, and central flow-through system.

PART II

«We cannot predict the future, but we can shape it.»
-Peter Drucker-

7 PLANNING A HEAT PUMP SYSTEM

While Part I of the book focused on pure knowledge transfer, this second part is about implementation. You'll be guided from planning a complete heat pump system, through its installation, and on to inspection and maintenance. Throughout, specific chapters with detailed explanations are referenced in the book to promptly answer any questions or deepen your understanding without significant effort on your part.

Planning a heat pump system lays the foundation for your home's heating supply. In this chapter, the individual steps for planning a heat pump system are sequentially listed and reinforced with brief examples. Figure 50 provides a recommended workflow for planning a heat pump system, known as the **planning scheme**, based on VDI 4645 guidelines. This planning scheme, also found in Appendix A, is central to the planning of a heat pump system. The specific steps for planning are detailed with explicit chapter references, ensuring you always have a clear point of reference. Each subchapter reintroduces the workflow to maintain clarity regarding the planning steps and progress. Please do not assume that you can go through the process in one go, step by step, and be done with it.

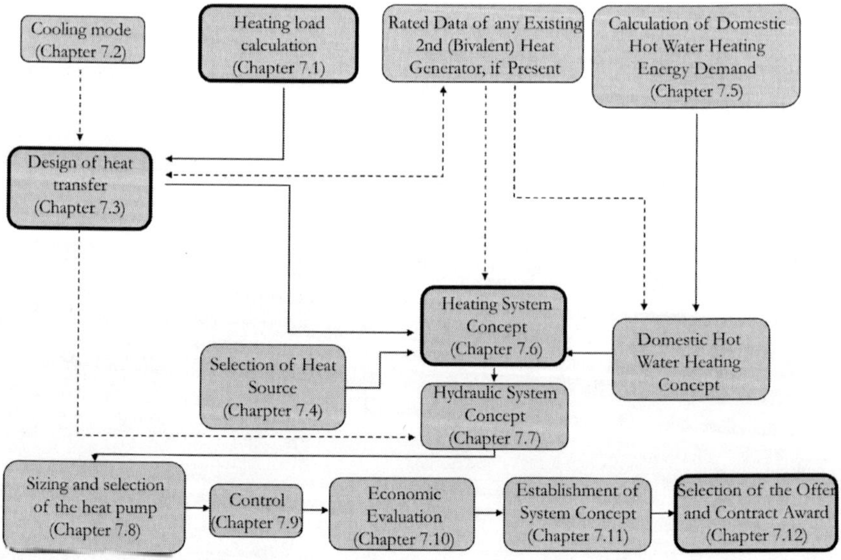

Figure 50: Step-By-Step Guide of a heat pump system, overall view

Rather, consider it as an iterative process. Based on new insights or a change in opinion, you might find it helpful to revisit and revise previous planning steps to devise the best solution for your needs. In Part III of the book, complete sample plans are presented, building on the knowledge acquired here.

Given that there are often questions about the responsibilities of the executing professions, Figure 51 provides an overview of the professional domains related to specific areas of expertise, using the example of brine and water heat pump systems.

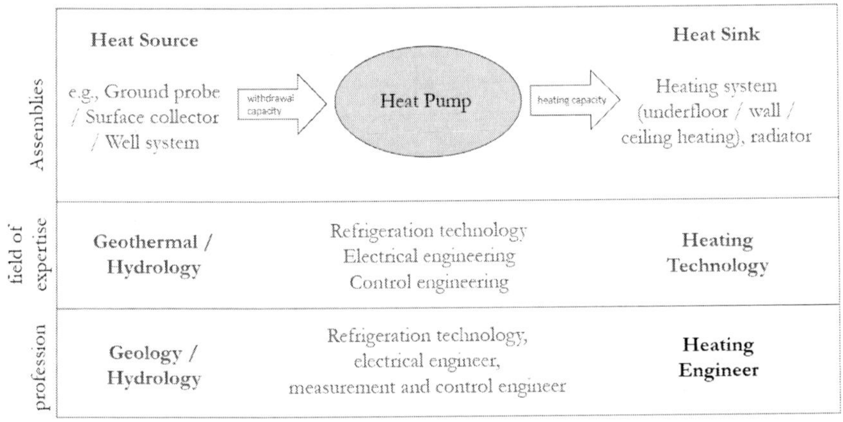

Distribution / Sales / Installation:
Heating contractor, electrical technician, drilling company, construction company, manufacturer, refrigeration technician

Figure 51: Overview of responsibilities in heat pump systems

7.1 Heating Load Calculation

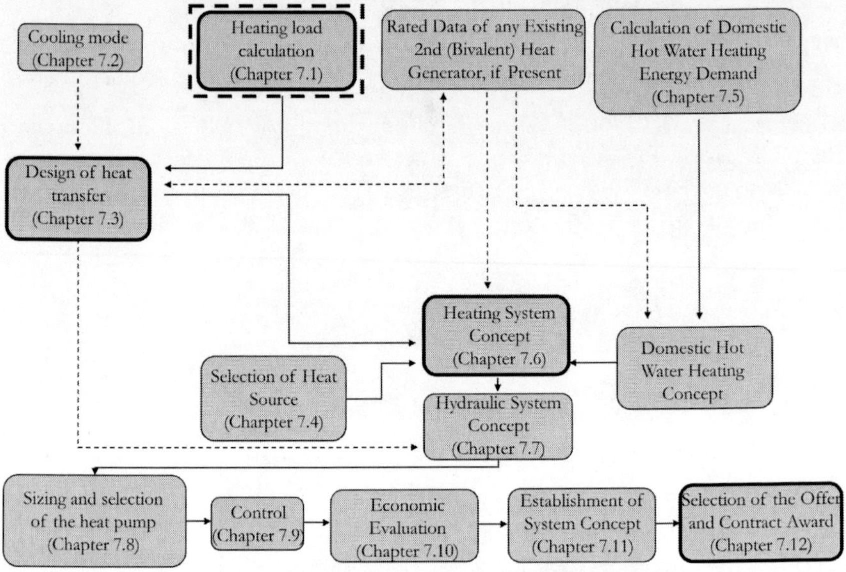

Figure 52: Planning scheme of a heat pump system, heating load calculation

The most crucial prerequisite for designing a heating system is determining the **heating load** ($\dot{Q}_{H,N}$). It is the starting point of the planning process. But what exactly is the heating load? To maintain a room temperature at a specific level, a balance between energy input and heat loss must be ensured. This means that the heating system must be capable of supplying the room with at least as much energy as is lost through heat loss in order to maintain a specific room temperature. If the goal is to increase the room temperature, the heating system must even be able to provide the room with more energy than what's lost due to heat loss. The heating load (also referred to as heating capacity) is given in watts (W) and indicates how much heat a building loses. The results of the heating load calculation provide insights on how to size the heat pump system. Therefore, the heating load is an essential parameter for an optimal sizing of the heat pump system.

In calculating the standard heating load, a distinction can be made between entire buildings or building units and individual rooms. There are also different methods for this, including **simplified methods** (two distinct procedures for buildings and rooms) and the **standard method** (which differentiates within the procedure between buildings and rooms). The simplified methods are limited to certain boundary conditions and applications, while the standard method does not have these restrictions. The following Table 3 from EN 12831-1 provides an overview of the calculation methods.

To calculate the heating load according to EN 12813-1 (please note the annexes), it requires the addition of the total transmission heat losses outside of the boundary area, **ventilation heat losses** if present and not explicitly excluded, **additional heating performance**, and **heat gains**.

	Procedure	**Application**	**Limitations regarding the overall scope of the standard**
1	Standard Procedure: Heat load of rooms, building units, and buildings	General approach for considerations on heat losses/heat load	
2	Simplified Procedure: Calculation of the standard heat load of a heated space (individual rooms)	Determination of room heat load before measures related to the heat emission system in individual rooms: — Replacement of the heat emission system — Hydraulic balancing	a) only applicable to spaces in 1) residential buildings or buildings with comparable use 2) existing buildings/building stock 3) buildings with natural ventilation
3	Simplified Procedure: Calculation of the standard heat load of a building	Determination of building heat load before measures related to heat generation, e.g., replacement of the heat generator	

a The simplified methods are not intended to replace the standard method in cases involving extensive measures related to generation and delivery.

Table 3: Calculation Procedure for Heating Load Calculation Overview

The calculation method is based on a stationary procedure. The boundary area can refer to either the entire building, the building unit, or the room. For a building, for example, it would be the sum of all transmission heat losses to the outside, whereas for a building unit, it would be the transmission heat losses to other building units.

Transmission heat losses refer to the losses through the structural components of the building envelope. The additional heating performance, for instance, is the capacity of an auxiliary heating system that's briefly provided after a heating break. This could be a bathroom heater fan, for example.

Significant factors influencing the heating load include the desired temperature, the thermal transmittance coefficients (U-value), the air exchange rate, and the building's storage mass.

The **desired temperature** is the indoor temperature that is used as the basis for calculating the heating load. You can either proceed according to standardized norm table values, or a targeted desired temperature can be agreed upon with the builder. To avoid disputes later on, the temperatures determined for the heating load calculation in consultation with the planning office should be documented in writing. At this point, it's important to note that with radiant heating systems, compared to convectors or radiators, a comfortable room temperature is achieved even with slightly lower air temperatures due to the increased radiant heat (by about 2-3°F). Even when determining the desired temperature, over-dimensioning of the heat pump system should be ruled out from the start. The thermal transmittance coefficient (U-value) indicates how much heat escapes through a square meter of the building envelope for every one-degree temperature difference. The air exchange rate denotes how often the volume of air in a room is replaced through ventilation. The more mass a building envelope has, the longer it retains the absorbed heat. This ensures that indoor temperatures remain stable over several daily cycles when outdoor temperatures fluctuate.

The dynamic relations between indoor and outdoor temperatures are simplistically viewed using a stationary calculation approach, which means that both the indoor and outdoor temperatures necessary for calculation are assumed to be constant. The indoor temperature depends on the specific use of the space, while the outdoor temperature is dependent on the location. The standard outdoor temperatures are based on test reference years.

Another simplification can be seen in not considering the internal heat entries and solar radiation.

It is important to note that oversized heat pumps operate less efficiently because they typically have short run times and high start-up frequencies. Conversely, if the heat pump is undersized, it cannot provide sufficient heat. From this alone, it is clear how crucial proper sizing of the heat pump is, which in turn makes determining the heating load essential.

It's not uncommon for existing heat pump systems to lack a heating load calculation based on the mentioned standards, resulting in the heat pump often being oversized under the assumption that "more is better." The reasons for a missing heating load calculation are evident. The calculation is complex and thus incurs corresponding costs at planning offices. These costs aren't usually compensated during the offer phase, and often, necessary documents, such as the thermal transmittance coefficient (U-value) of the building envelope, are missing during new construction proposals. For an existing building, especially older ones, this can be even more complicated. In cases where the homeowner isn't the original builder, blueprints, U-value specifications, and, above all, details about the specific structure of individual layers of the building envelope (e.g., interior plaster, masonry, insulation, and exterior plaster) are frequently missing. To calculate the U-value in such cases, there must be an accurate determination of the materials and thicknesses of specific building envelope components (e.g., walls). Measuring the U values with a measuring device is also conceivable, but due to the high effort involved, it's rarely implemented in practice. This demonstrates that, especially for such existing buildings, a certain degree of uncertainty can be factored into the calculated heating load. To validate the calculated heating load, comparisons can be made with the fuel consumption from previous years. However, these consumptions usually include both the building's consumption and the hot water preparation consumption. Furthermore, the number of heating days can lead to significant deviations.

Even if no explicit heating load calculation was carried out during the offer phase, it is essential to conduct this in accordance with the mentioned standards using a detailed procedure once the order is received. The **approximate determination of the heating load** can be used for preliminary planning (see Part 3 of the book). For new constructions, this is straightforward anyway, and for some energy certificates, the heating load is already specified.

7.2 Cooling Mode

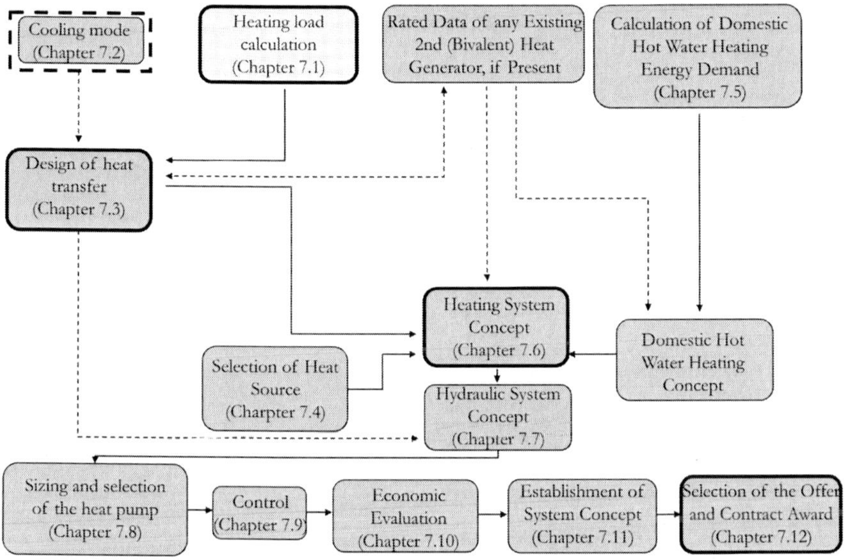

Figure 53: Planning scheme of a heat pump system, cooling mode

Having determined the heating load (Chapter 7.1), the future heat pump operator should ask themselves whether they want to use their heat pump for cooling during the warmer months. Especially now, with global warming being a prevalent topic and the expectation of increasingly longer and intensifying heating periods, the aspect of cooling should not be overlooked.

The use of heat pumps for cooling is explained in more detail in Chapter 3.3.

It becomes clear that, when cooling with a heat pump, one should differentiate between the type of cold transfer into **silent (static)** and **circulated (dynamic) cooling**. Furthermore, there is a need to distinguish the cooling process technology, which means **active** and **passive cooling**.

Initially, the user should determine their personal requirements for room temperature, humidity, and thermal comfort. Thermal comfort is not uniformly measurable due to the individual perception and adaptability of each person. Determining it involves considering room air temperature, wall surface temperature, relative humidity, and room air circulation.

Once these requirements are set, the building cooling load is determined. In contrast, for heating, the heating load is determined (see Chapter 7.1). The VDI 2078 can be used to determine the building cooling load.

Having determined the building cooling load, one then decides whether to install active or passive cooling (see Chapter 3.3 for reference). In general, passive cooling is preferable, as active cooling usually requires higher energy consumption.

The building cooling load should then be compared to the cooling capacity of the heat pump. In addition, the design of the heat source must be reviewed regarding its cooling function dimensioning, potentially leading to necessary adjustments. In general, the system should operate efficiently in both heating and cooling modes.

Once all these steps have been completed, the technical implementation of the system for explicit cooling function is required.

7.3 Design of Heat Transfer

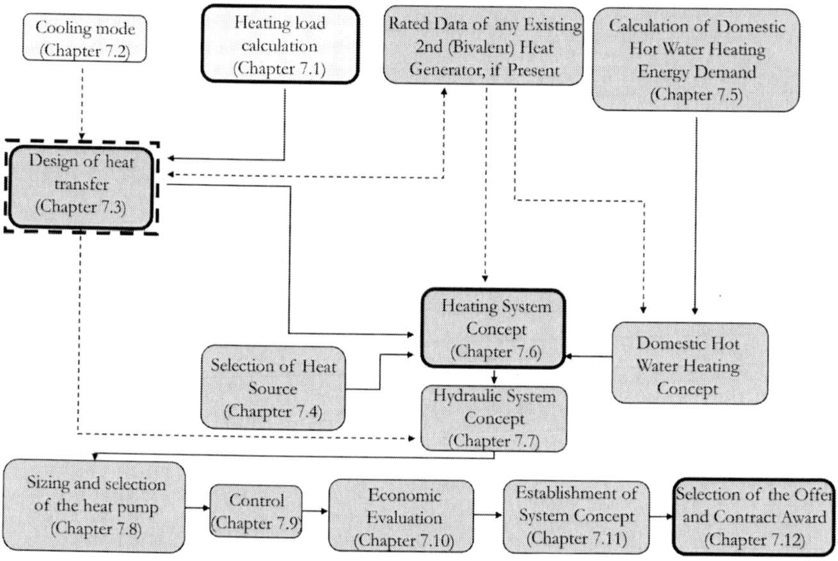

Figure 54: Planning scheme of a heat pump system, design of heat transfer

We are already familiar with the building's heat load (Chapter 7.1). We also know whether we want to use the heat pump for cooling during the warm season (Chapter 3.3 and Chapter 7.2). Now the question arises about the proper planning of the **heating surfaces**.

In the planning of the heating surfaces of a heat pump system, most mistakes are made based on experience. Initially, when choosing the heating surfaces, it should be clear whether they are used for heating or also for cooling. Even in new constructions, where a heat pump system is not yet planned, potential future installation of a heat pump should be considered, and the heating surfaces should be sized accordingly.

A decisive parameter for the efficiency of the heat pump is the system temperature of the heat utilization system. The system temperature of the heat utilization system consists of the flow and return temperature. Both of the following points are crucial for the heat pump's coefficient of performance and thus the operating costs. A maximally high coefficient of performance is achieved through:

- The smallest possible temperature difference between the heat source and the heating medium.
- The lowest possible system temperature.

The coefficient of performance indicates the ratio of benefit to effort, where the benefit is defined as the heating output and the effort as the compressor's drive power.

To be able to set the lowest possible flow temperature, surface heating systems are particularly suitable for saving energy. Floor, wall, and ceiling heating systems are referred to as surface heating systems.

Due to the high proportion of radiant heat, the perceived temperature with surface heating systems is 35 to 37°F higher than the perceived temperature when heating with radiators. This, in turn, means that the room temperature with surface heating systems can be set 35 to 37°F lower than with radiators. Thus, flow temperatures of 95°F can be achieved with surface heating systems. To keep the flow temperature as low as possible, the area covered with underfloor heating should be as large as possible, which means it should also be installed under showers and in bathrooms. Furthermore, surface heating systems are almost invisible since they are space-savingly built into the floor, wall, or ceiling. Surface heating systems can be implemented electrically or in the form of pipes filled with water. The connection to heat pumps is explained in more detail below, focusing exclusively on water-based surface heating systems.

The **underfloor heating** system is probably the best-known type of surface heating. With underfloor heating, coiled tubes filled with warm water are laid under the screed. The water distributes the heat across the floor. While the conventional convector primarily heats the air, thus creating circulation in the room (convective heat), underfloor heating systems warm from below with pleasantly evenly distributed radiant heat. The room climate, often perceived as uncomfortable due to dry air or stirred-up dust with convective heat, is not an issue with underfloor heating or any surface heating system. Most of the heat is at the floor, warming it entirely, which in turn serves as a heat store. When the heating thermostat is turned up, the heat first goes into the screed, then through the floor covering into the room, resulting in a rather sluggish heating behavior.

Due to the heating of the floor, the underfloor heating system can compensate for short heating breaks and ensures continued heating through the already warmed floor. With underfloor heating, a distinction is made between the **dry construction** method and the **wet system**, with the names referring to the type of installation. The dry system is usually suitable for renovation measures. There is no need to cut into the screed; instead, the installation takes place on top of it. After verifying that the subfloor is suitable for the procedure and has the necessary load-bearing capacity (a load distributor layer may be required), an insulation layer with edge insulation strips is laid on it (system panels). This ensures that there is no heat loss downwards or to the sides. Heating pipes are laid in the system panels, and the system is finished off with a floor covering of choice on top. Due to its compact properties, the entire system has a low installation height, so it's not noticeable in terms of appearance or space. With the wet system, in contrast to the dry system, the screed is poured over the installed heating pipes or embedded in the still wet screed. The direct contact between the pipes and the screed in the wet system results in a much more efficient heat transfer compared to the dry construction method. Furthermore, in the wet system, the screed heats up, unlike in the dry construction method, and thus serves as a storage medium for further heat release.

A common misconception is that underfloor heating must always produce a warm floor. This should not be the case if the underfloor heating is used energetically correctly. Instead, the underfloor heating covers the building's heat loss and achieves the desired room temperature. Once the desired room temperature is reached, the thermal actuator receives a signal from the room controller to shut off the heating.

1 Substructure; 2 Panels; 3 Pipe; 4 Filling material; 5 Flooring

Figure 55: Underfloor heating in dry construction

The heating circuit remains closed until the desired room temperature is once again reached. If one wishes, contrary to energy principles, to achieve a constantly noticeably warm floor, the room thermostat should be set to the maximum value, the actuator should be manually opened, or a linearly operating actuator should be used.

However, this approach is not recommended for energy conservation. Regardless, the self-regulating effect of the underfloor heating persists, and the heat is delivered to the rooms in a controlled manner.

Figure 55 and Figure 56 depict underfloor heating in dry construction and wet systems.

Thermal **concrete core activation** (also known as concrete core tempering or thermal component activation) is a special form of underfloor heating, in which the heating pipes are directly embedded in the floor or ceiling slab.

In this method, the heating pipes are directly cast into the floor or ceiling slabs. This approach, differing from traditional underfloor heating, results in an even more effective utilization of the building's mass, and thus, its heat storage capacity.

Figure 56: Underfloor heating in a wet system

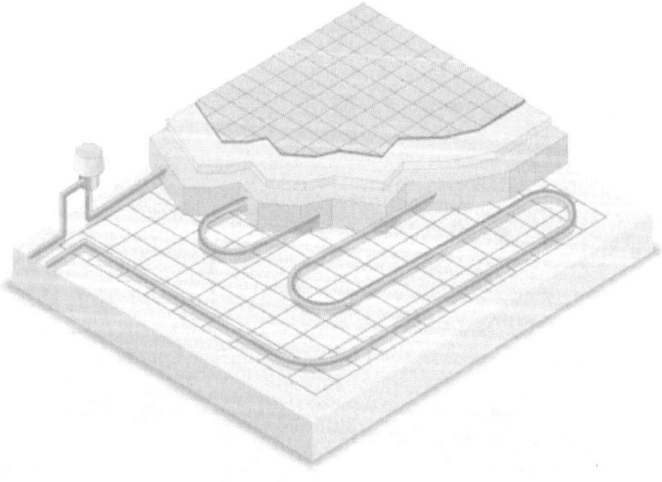

Figure 57: Principle of concrete core activation

In this way, solar fluctuations can be very effectively compensated for. Just like with underfloor heating, thermal energy is transferred to the components, which then radiate it upwards or downwards. The inertia of the heating system is naturally somewhat greater here than it already is with traditional underfloor heating. For this reason, programs such as nighttime temperature reduction are only marginally effective with thermal concrete core activation. This type of radiant heating, thanks to its massive heating surfaces, operates with the lowest supply temperature. Figure 57 illustrates the principle of thermal concrete core activation.

With **wall heating** (see Figure 58), similar to underfloor heating, the component is warmed by a pipe system in the wall through which water flows. Here too, a distinction is made between dry and wet construction methods. Wall heating can be installed on both the exterior and interior walls, but when installed on an exterior wall, impeccable insulation is required. If wall heating is installed on interior walls with poorly insulated exterior walls, it can lead to unpleasant drafts. Due to insulation requirements, wall heating systems are generally best suited for new buildings or low-energy homes that are ideally insulated. One aspect that must not be overlooked with wall heating is that furnishing should be avoided on the heated walls.

Figure 58: Wall heating "ACULAR-M" by THERMOLUTZ GmbH & Co. Heizungstechnik KG

The risk of heating performance being obstructed by furniture doesn't exist with **ceiling heating**. Here, a distinction is also made between dry and wet construction methods. In the dry construction method, gypsum fiber boards with integrated heating pipes (see Figure 59) are typically attached to a substructure on the ceiling. However, it can also be implemented separately above the gypsum fiber boards. In the wet system, the heating pipes are directly covered with plaster. The ceiling heating also radiates its heat, rather than using convection. Since warm air rises, a "heat cushion" forms in tall room areas, which can lead to inefficiencies due to the positioning of the heat output. Such "heat cushions", however, can be minimized with thoughtful planning, especially in new constructions. Ceiling heating systems are advantageous, for instance, during renovation of old buildings where the floor is to be preserved, or when one doesn't want to select floor coverings based on their suitability for underfloor heating in a new construction.

The aforementioned radiant heating systems can also be combined as desired. For example, it's not uncommon for buildings to have several rooms heated with underfloor heating and a few spaces heated with wall or ceiling heating.

Figure 59: Ceiling heating and cooling modules "SYSTEM acular DECKE" by THERMOLUTZ GmbH & Co. Heizungstechnik KG

This can be considered when rooms with wall or ceiling heating are equipped with a floor covering that has an unfavorable (high) thermal resistance, and the owner still doesn't want to forgo that particular flooring. For instance, in a living room with underfloor heating that's laid with parquet, a few additional square meters of wall or ceiling heating might be installed, allowing for heating at a desirable low flow temperature of 95°F. Conversely, the additional heating surfaces can also be avoided by either raising the flow temperature at the expense of efficiency, or by choosing a floor covering with a more favorable (lower) thermal resistance.

Let's delve deeper into the concept of **thermal resistance** (R). Thermal resistance is the inverse of the thermal transmittance coefficient (U) and should be as low as possible for materials covering the heating system, such as floor coverings. The thermal resistance is measured in the unit m^2K/W, and the limit that should never be exceeded for a floor covering used with underfloor heating is $0.15m^2K/W$. The thermal resistance is abbreviated as R. It is calculated by dividing the thickness of the floor covering "d" (in meters) by its specific thermal conductivity "λ" (in W/mK). If the floor covering consists of multiple layers, the values must be summed up. The computed thermal resistance provides insights into the heating effect within a room. If the thermal resistance is very high, it takes longer for the heat to penetrate the floor covering and warm the room, leading to reduced efficiency. Conversely, with a low thermal resistance, the heat quickly permeates through the floor covering into the room, resulting in rapid heating without significant efficiency losses. Independent calculation of the thermal resistance of the chosen floor covering is redundant since manufacturers typically provide this information. For underfloor heating, ceramic tiles or natural stone are ideal due to their very low thermal resistance. Wooden planks have a high thermal resistance, making them less suitable. Parquet is better in this regard, but it is still far from the low thermal resistance of a ceramic tile. Laminates should be avoided with underfloor heating, and for carpets, there are specific types designed for use with underfloor heating, indicated with a manufacturer's pictogram (see Figure 60).

Figure 60: Carpet pictogram: Suitable for underfloor heating floors

Table 4 provides an overview of the thermal resistance values of common floor coverings.

It is essential, during the planning phase of the heat pump system, to meticulously focus on an energy-efficient solution and have the planner conduct a detailed calculation. This includes the precise selection of the floor covering when using underfloor heating. Often, for existing heat pump systems, precise calculations are lacking, resulting in the flow temperatures of the underfloor heating reaching 113°F or the panel radiators reaching 122°F. If only one room in the building is heated at a higher flow temperature compared to the others, this can lead to a significant deterioration in the coefficient of performance. All rooms in the building must always be heated at the same, and ideally lower, flow temperature because the heat pump controller must always be set to the necessary maximum heating curve. If flow temperatures are defined differently by room, the heating curve would always need to be adjusted to the higher flow temperature, leading to inefficient operation.

So, how should one approach the installation of a **heat pump system in existing** heating setups? It always depends on how professionally the planning of the existing heating system was carried out in the past. If done professionally, optimization can proceed without obstacles, and straightforward measures can usually be taken. However, if the planning of the existing heating system was less specialized, the amount of work, and the associated costs, could significantly increase.

Material	Thickness in mm	Thermal resistance (R) in m²K/W
Ceramic tiles	13.00	0.012
Cork parquet	11.00	0.129
Wood parquet	12.00	0.125
Wooden planks	17.00	0.130
Plastic PVC	2.00	0.010
Vinyl plastic	3.00	0.015
Laminate	9.00	0.044
Linoleum	2.50	0.015
Marble slabs	30.00	0.014
Mosaic parquet	8.00	0.038
Natural stone	20.00	0.017
Needle felt	6.50	0.120
Strip parquet	22.00	0.105
Carpet	8.00	0.100

Table 4: Thermal Resistance of Common Flooring Materials with Thicknesses

Achieving a low supply temperature often becomes somewhat complex.

Plans and calculations for the heating surfaces usually haven't been conducted and therefore aren't available. As a result, high supply temperatures and significant power consumption typically occur.

The notion of rapidly heating rooms should generally be discarded. This would only lead to high supply temperatures, thus reducing efficiency. Instead, the focus should be on deliberately slow heating to incorporate the building's storage behavior, which in turn requires lower supply temperatures.

And this is precisely where adjustments need to be made, namely in **reducing the supply temperatures**. Several options for this will be described in the following sections.

Firstly, depending on the construction year of the existing building, the building envelope should be examined. In some cases, there might not be a need for extensive insulation of the outer walls; simply replacing windows and doors might suffice. Additionally, attention should be given to insulating the basement ceiling and the ceiling adjacent to the attic. If the roof is insulated and the area directly beneath the roof is in use, insulation of the topmost ceiling can be omitted. To plan the necessary insulation in the existing building, it's crucial not to forgo consulting a qualified energy consultant. The energy consultant can provide an energy certificate and will highlight where the greatest energy-saving potential exists in the building in combination with the lowest costs. Under no circumstances should one opt out of consulting a qualified energy consultant, as incorrectly implemented insulation measures can lead to excessive condensation within walls, subsequently promoting mold growth. Especially with interior insulations, which often seem more practical to laypeople, condensation within the building envelope is promoted. It's not uncommon to encounter, during the demolition of an old building that underwent various renovation measures over the years, that the inner parts of the outer walls are widely affected by mold. On top of that, the property owner may have been suffering from respiratory diseases for years. By implementing targeted insulation, heat loss is reduced, which in turn allows for a reduction in the supply temperature.

In the second step, the heating surfaces are assessed based on the previously mentioned aspects. If necessary, for instance, a replacement of the floor coverings can be done, or an addition of a wall heating system can be implemented, so that the underfloor heating operates more efficiently and can again be run with a lower supply temperature.

In the next step, a desired minimum temperature for each room should be determined, and in the rooms to be heated, all heating circuits and thermostats should be fully opened. The warmest room in the building is referred to as the reference room, which is typically either the living room or bathroom. Subsequently, the supply temperature is gradually reduced so that the desired minimum temperature in the reference room is just about achieved. If a supply temperature of 113°F is still required, there's a need in the rooms for an addition of heating surfaces in combination with the old heating surfaces or a complete renewal. When adding heating surfaces, a pressure loss calculation is necessary as well as the execution of the hydraulic balancing (see Chapter 8.5).

7.4 Selection of the Heat Source

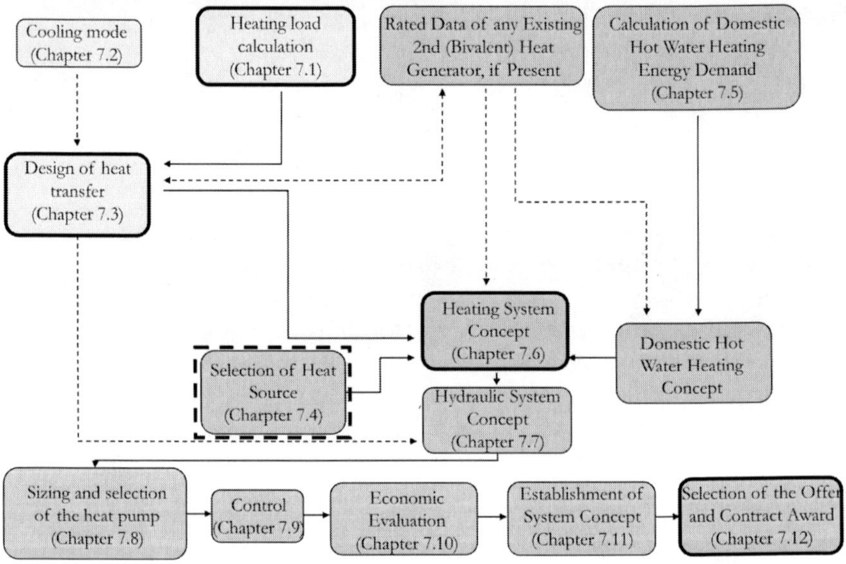

Figure 61: Step-By-Step Guide of a heat pump system, selection of the heat source

Having now understood the design of the heat transfer (Chapter 7.3) based on the determined heating load (Chapter 7.1) and the potential cooling operation (Chapter 7.2), a decision is needed regarding the explicitly suitable heat source.

The selection of the heat source should be based on **investment costs, operating costs**, and the **availability** of the heat source.

Investment costs are primarily determined by the development costs for the heat source. In the case of ground and water heat sources, extensive heat source investigations are required. Additionally, there are costs for the heat pump system tailored to the specific heat source.

Let's first consider **availability**.

For the **air heat source**, the stark advantage immediately becomes apparent: there are no restrictions concerning availability. Air is always and everywhere present on any property. Therefore, there is considerably less to consider with the air heat source than with ground and water sources, which explains why air-source heat pumps are predominantly offered nationwide by all-in-one providers. A major focus with the air heat source, and the associated air-source heat pumps, is the operational limit, which depends on the technology, the specific refrigerant used, the location, and the installed components. Depending on these factors, the heat pump can be deployed at lower or higher sub-zero temperatures. For this reason, a heat pump with, for instance, a lower operational limit of 5°F should not be operated in a location with outside temperatures down to -4°F. Current discussions about global warming should not be optimistically factored in at this juncture, because even if temperatures become milder over the years, the lowest outside temperature remains crucial for the air-source heat pump. It's better to be on the safe side, as assuming there won't be lower future external temperatures could be a risky bet.

If the temperature falls outside the operational limit of the air-source heat pump, it will shut down and only restart once the operational limit is again surpassed. Below the operational limit, if installed, the electric heating element is automatically activated. Most energy providers (network operators) typically limit the electric heating element to 1.5 times its nominal electrical output.

Moving on to the water heat source. The thermal use of the subsoil is regulated in the VDI 4640 Part 2. This also includes the utilization of groundwater with well systems. As already described in Chapter 4, water is the best heat carrier. If one intends to use water as a heat source, the planning, construction, and operation must comply with the specific water and mining regulations, as well as the explicit state-specific rules. The intent with the water heat source is primarily to utilize near-surface groundwater. If there's an intention to use deeper groundwater as a heat source, special protective measures are necessary. In general, the drilling of the well system is rather complex and should be undertaken by qualified planning offices. Initial planning errors, disproportionate cost-cutting measures, or even overlooked regulations can end up being costly in hindsight and should be avoided at all costs. However, even if there are a multitude of regulations and one might discuss the standardization or consolidation of these, they fundamentally serve a uniform purpose:

- The drinking water supply for the population takes precedence over the thermal use of groundwater.
- Substances that pose a risk to groundwater must not enter it.

- Careful handling of groundwater should be ensured.

From the nominal flow rate, the sizing of the supply and discharge wells is determined. It is essential to note that the yield of a well depends on the local geological-hydrogeological conditions. This must be proven through pumping tests. For example, if there are vast distances between the ground surface and the groundwater level, this implies an increased pump performance, which in turn negatively affects the efficiency of the heat pump system. To still utilize the water heat source with the increased electricity demand of the supply pump, a photovoltaic system for self-consumption could be deployed. As a result, after an economic efficiency analysis, economical utilization can be achieved again. Besides the availability of the suction well, it is also essential to ensure the uninterrupted intake of the pumped water volume using a discharge well. Furthermore, the groundwater flow direction must be considered so that neighboring well systems do not influence one's system. A possible "worst-case" scenario here would be that the neighboring well pump system, with the cooler water to be returned, affects one's suction well due to an unfavorable flow direction, resulting in its constant cooling.

Sometimes, when planning such a well system, there's talk of a numerical hydraulic-thermal groundwater modeling. Such a model serves to identify mutual influences of neighboring systems and to exclude a hydraulic short circuit. Please don't be deterred from the water heat source because of this, as it is truly recommended only for larger installations that go beyond individual home use. Otherwise, it would only be a cost driver that exceeds proportionality. It's essential that there is a suitable groundwater mechanism in place for using the water heat source. Only then can long-term trouble-free operation be guaranteed. This means that the risk of corrosion of metal and concrete components and iron staining (or "ochering") must be excluded. Corrosion can be promoted by the hydrochemical composition of the groundwater. Ochering refers to the precipitation of iron hydroxides and manganese oxides in water-operated systems. Over time, this leads to the "clogging" of water-carrying and water-pumping components. From all the mentioned scenarios and requirements, the need for a hydrogeological expert opinion arises.

If you are considering such a well system, you should initially choose a location on the property that is easily accessible for maintenance work. Also, the area of the well shaft should not be close to traffic areas like garage driveways.

The **heat source from the Earth** can also be tapped relatively independent of external temperatures but is also subject to a permitting process.

If the regeneration time of the ground for **geothermal probes** is too short, the temperature in the ground can decrease permanently over time. This is not permissible, and according to VDI 4640 Part 2, which also applies to the soil itself as a heat source, is regulated. VDI 4640 Part 2 states: "The design of geothermal probes must be such that for a given load profile, the fluid and probe temperatures during the planned operating period... remain within specified limits." But how can there be such a permanent drop in the brine temperature or an insufficient regeneration time? The geothermal probe extracts heat from the ground over a defined period (see Chapter 4.2.1). When this process is periodically repeated, for example, year by year, a balance is achieved. In this equilibrium, the thermal energy applied over a period is compensated for by a heat exchange with the surrounding environment. The environment includes the Earth's surface, lateral ground areas, and deeper subterranean zones. From this surrounding, the extracted heat is compensated for. However, if the ground probes are undersized and too much heat is withdrawn, the required compensation cannot occur, and the ground cools down. Thus, in this case, an insufficient sizing of the geothermal probe leads to an undesired situation.

To investigate the precise thermal properties of a substrate concerning its heat utilization, a **Thermal-Response-Test (TRT)** is conducted. This method determines the effective thermal conductivity of the ground over the entire length of the probe as well as the thermal borehole resistance, with the focus being on the theoretical approach of the Kelvin line source model. Thus, the TRT does not directly determine a specific extraction power, but instead measures the thermal properties of the substrate. The thermal properties obtained from the TRT form the basis for determining the extraction power, which is determined with these values in another design process. This method thus identifies whether the location is suitable at all and, if so, how the geothermal probes need to be dimensioned. In general, the TRT can also be used to detect errors or for performance control. The Thermal-Response-Test (TRT) is often also referred to as the Geothermal-Response-Test (GRT). The deeper theoretical foundations, the experimental setup of a measurement system, the execution of the measurement, and the evaluation of the TRT measurement for geothermal probes are covered in VDI 4640 Part 5.

For **surface collectors**, the regeneration time is not considered critical if they are installed according to the specifications of VDI 4640 Part 2. What's interesting about the surface collector is that it can often be laid as a DIY project. Since surface collectors are not located very deep in the ground, it's important to ensure that they aren't built over later on. In general, the surface should not be sealed, and there must be a distance of at least 1 meter from buildings and foundations. The installation depth of 1.5 meters must not be exceeded due to the surface-dependent function of the surface collectors compared to the geothermal probes. While higher depths have a higher temperature, the connection to the earth's surface and thus the regeneration is lower at those depths. Specifically, the surface collector should be laid at a depth of 1.2-1.5 meters and be at least 30 cm below the frost line. Moreover, a greater installation depth would mean additional effort.

It doesn't hurt to choose a slightly larger laying distance if possible. While an oversized heat source of the surface collectors might be a bit more expensive, this will pay off over time through the higher brine temperature and the associated improved coefficient of performance (COP) of the heat pump. On the other hand, if the heat source of the surface collectors is undersized, the brine temperature might show too low a negative range during a very cold heating period, causing the heat pump to shut off or the COP of the heat pump to decrease.

In addition to the geothermal probe and surface collectors, chapter 4.2 also discussed special forms like the **trench collector**, **geothermal baskets**, **borehole piles, short geothermal probes**, **trench collectors**, etc.

For **geothermal baskets**, a distinction is made between cylindrical and conical designs (see chapter 4.2.3.2). With the conical design, the extraction power is calculated based on an adapted cylindrical design.

For **borehole piles**, the energetic design is performed analogously to the ground heat exchangers (also known as geothermal probes). However, different operating conditions apply to the borehole piles. Thus, the calculation method applicable for geothermal probes is only valid as long as the ratio of length to diameter of the borehole piles is sufficiently large (≥ 20:1). Furthermore, the pile diameter must not exceed 50cm.

It's crucial to note that the performance and design values applicable to probes cannot be directly transferred to borehole piles. The borehole piles, which simultaneously ensure static load transfer, must never reach or fall below temperatures in the frost range, as this could endanger the load transfer. Therefore, the design must be carried out accordingly so that the heat transfer fluid, at the entry into the earth-contacted concrete components, never falls below the frost range. This must be ensured in operation by a safety device and also a safety margin of 2 degrees Fahrenheit. The permissible maximum temperature, dependent on the materials used and the maximum permissible temperature gradients, must be coordinated with the authorities.

For **short geothermal probes**, identical requirements as for surface collectors must be observed during installation. However, with the short geothermal probes, a distance of at least 2m to buildings, foundations, or similar must be maintained.

For **trench collectors**, a design table is also listed in VDI 4640 Part 2. However, due to the different installation options and the associated installation complexity, it is recommended that the determination of the extraction power be carried out using special software by a specialist planner.

Table 5 provides a general overview of the different heat sources and aims to assist you in making a selection. For the heat source "Sole" (geothermal energy), a distinction is made between the geothermal probe and the geothermal collector. For a description of the cooling process, please refer to Chapter 3.3.

In this context, Appendix C, the **heat pump matrix**, will be helpful in selecting the heat source.

Heat source	Efficiency	Investment costs	Heat pump installation location	Cooling
Water	best	high	inside	passive
Ground probe	very good	very high	inside	passive
Ground collector	very good	high	inside	not possible
Air	good	low	outside or inside	active

Table 5: Selection Scheme Based on Heat Sources

7.5 Determination of Energy Requirements for Domestic hot Water Heating

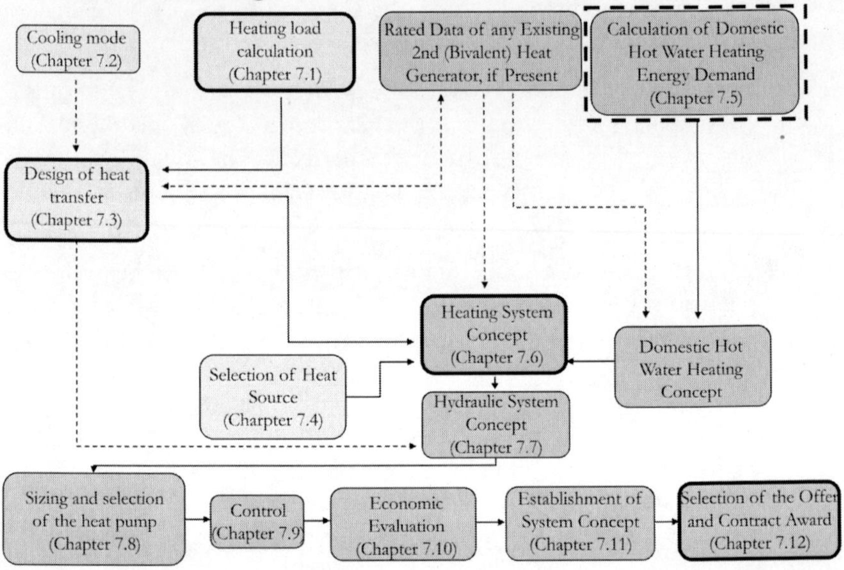

Figure 62: Step-By-Step Guide of a heat pump system, determination of hot water heating energy demand

When considering meeting the heating demand (see Chapter 3.1) in the house by upgrading to a heat pump, the question of implementing domestic hot water heating (hot water demand; see Chapter 3.2) with the heat pump naturally arises. It's important to note that as the standard for building insulation improves, the proportion used for space heating steadily decreases. However, the portion for domestic hot water heating remains constant.

It's wise to roughly estimate 0.15kW per person in the household for hot water preparation. This approach takes into account storage losses, distribution losses, and typical load profiles.

In general, it can be said that the hot water demand can vary significantly from one household to another. Therefore, it's essential to truly determine the individual habits of the new heat pump owner when it comes to using hot water. Is showering more common, or does the household tend to bathe? How large is the bathtub, and are rain showers used when showering? Just the design of the bathtub, whether it's a standard or triangular tub, is relevant for the given water volumes. If the bathtub holds, for example, 200 liters, then a 170-liter hot water tank would not be sufficient. Showering with a rain shower head also requires larger amounts of water.

Once all these parameters for the household have been determined, they should be recorded, and from this, the total hot water demand (domestic hot water demand) should be calculated (also refer to Chapter 6).

When heating domestic water with a heat pump, lower supply temperatures are available compared to water heaters powered by oil or gas. This means that the indirect hot water storage tank of the heat pump typically requires a larger heat exchanger surface than the water tank heated with oil or gas. For this reason, during renovations, the old water heater of the conventional heating system must be replaced by a water heater with a larger heating surface. For each KW of heating power, about $0.25m^2$ of heat exchanger surface is required.

In planning the hot water storage tank, the existing water quality must also be taken into account, so that if necessary, a water treatment system can be included in the design. If the local water quality is unknown, a water analysis should be conducted. More detailed information on hot water storage tanks and domestic water heating can be found in Chapter 6. Based on Chapter 6, the **concept for domestic water heating** can be developed.

7.6 Heating Concept

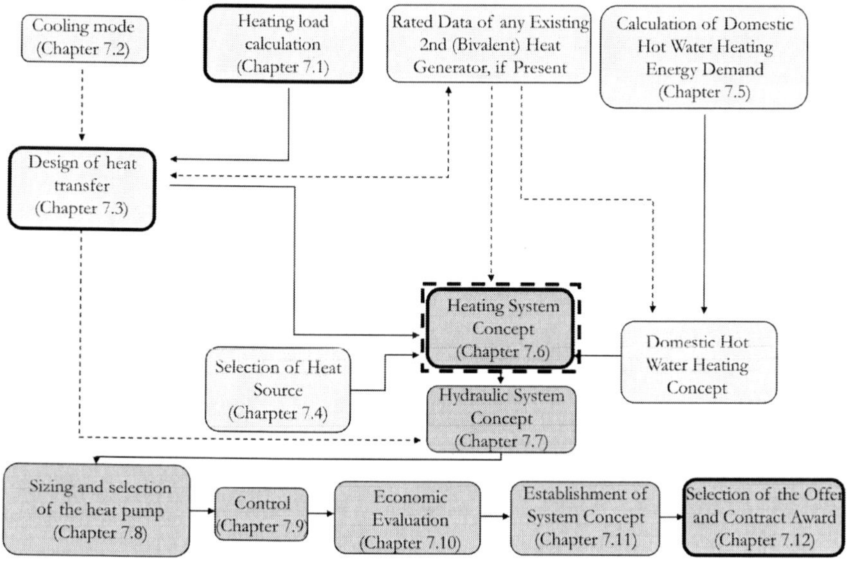

Figure 63: Step-By-Step Guide of a heat pump system, heating concept

Based on the determined foundations, a conceptual design of how the heat pump system should fundamentally look can now be created. The concept can be developed based on individual preliminary investigations and collected information from the previous planning steps (Chapters 7.1 to 7.5). The heating load of the building, the energy requirement for domestic hot water heating, and the associated domestic hot water heating concept, as well as the type of heat source, have already been determined and defined. The decision on whether a second heat generator will be used for mono-energetic or bivalent operation (see Chapter 5) should also be incorporated into the heating concept design. Furthermore, the decision on whether cooling operation via the heat pump should be ensured needs to be made. Depending on the given conditions, it's now possible to set up several system concepts and then plan each concept in detail in the subsequent chapters, comparing their cost-effectiveness (see Chapter 7.10). Appendix B lists the **planning checklist. With this, the foundation for the upcoming detailed planning has been laid.**

7.7 Hydraulic Concept

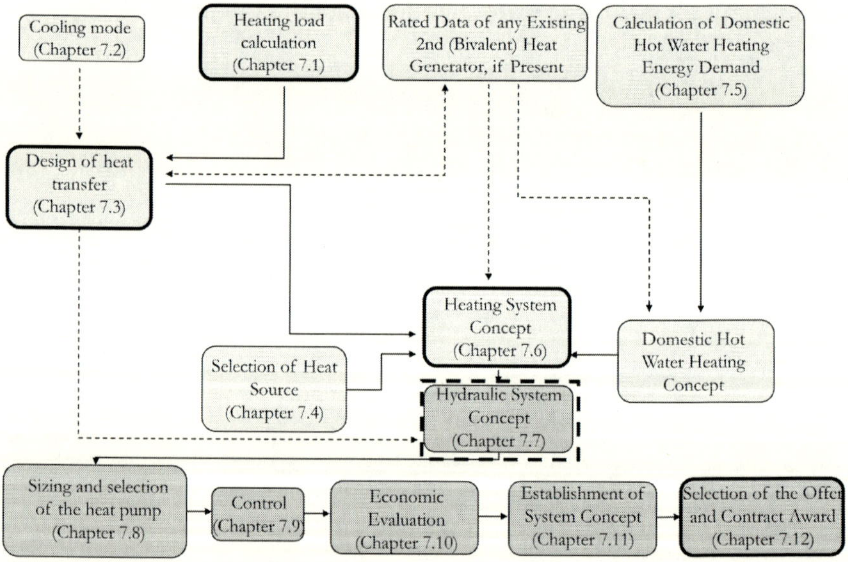

Figure 64: Step-By-Step Guide of a heat pump system, hydraulic concept

With the various heating concepts now available (see Chapter 7.6), the question arises regarding the suitable hydraulic concept. Depending on the desired application, there are various hydraulic concepts. An essential component of the hydraulic concept is the selection and implementation of the buffer storage (see also Chapter 6).

Once a type of buffer storage has been decided upon (see Chapter 6), the design of the chosen buffer storage follows. The design of the buffer storage is differentiated into bridging shut-off periods, optimizing runtime, or providing the minimum volume in the distribution pipes (detailed explanations of these functions can be found in Chapter 6.1).

For the **buffer storage used to bridge shut-off periods**, it must provide the required water volume from its storage during the time the heat pump is not in operation. If radiators or surface heating dry systems are installed in the house, a buffer storage is necessary.

However, it is not necessary to design the buffer storage based on the entire energy volume of the shut-off period in the building. As a rule of thumb, in residential buildings with a maximum of 2 hours of shut-off time, a value of **60 liters of buffer storage volume per kilowatt of heating** capacity can be assumed.

If wet floor heating systems are installed in the house, the storage mass of the screed or wall is sufficient to bridge the shut-off times.

If the buffer tank is used for **runtime optimization**, the intention is to optimize the heat pump for the longest possible runtime with extended pauses in operation, regulated by the heat pump controller. In the case of radiators, a buffer tank is also required for runtime optimization due to the lower water content and lower mass. Surface heating systems are typically designed without a buffer tank for runtime optimization due to their minimum volume. It is recommended to have a buffer tank volume of **20 liters per kilowatt of heating capacity** for this purpose.

To ensure the safe operation of the heat pump, buffer tanks are used to **provide the minimum volume** required for safe heat pump operation. The minimum volume is **3 liters per kilowatt of heating capacity**. In heating circuits with static heating surfaces, the minimum volume must be available even without heat demand (thermostat closed). For underfloor heating systems, the minimum volume must be provided for the pipes from the heat generator to the distributor or system separation.

However, as previously mentioned, with floor heating, the minimum flow rate can also be ensured through the heat distribution system, which ultimately may lead to the installation of a **heat pump system without buffer storage**. The minimum flow rate is taken from the technical data. The flow rates of the heating circuits are derived from the heating surface calculation. This way, the function of the buffer storage can be replaced by unregulated heating circuits or a hydraulic interface. Often, just three to four unregulated heating circuits are sufficient to meet the condition, usually using the heating circuits of the bathrooms and constantly heated rooms. According to the Energy Saving Ordinance, it is stipulated that each room must be subject to individual control. This would contradict the aforementioned approach. An application for exemption from individual room control can be made to allow for the self-regulation effect of the underfloor heating and the minimal cycling of the heat pump.

An important principle to consider during planning, based on experience, is "As simple as possible". Complex systems often lead to complex problems and are generally more prone to malfunctions. Therefore, do yourself a favor and stick to the simple and common hydraulic systems when planning heat pump installations.

7.8 Sizing and Selection of the Heat Pump

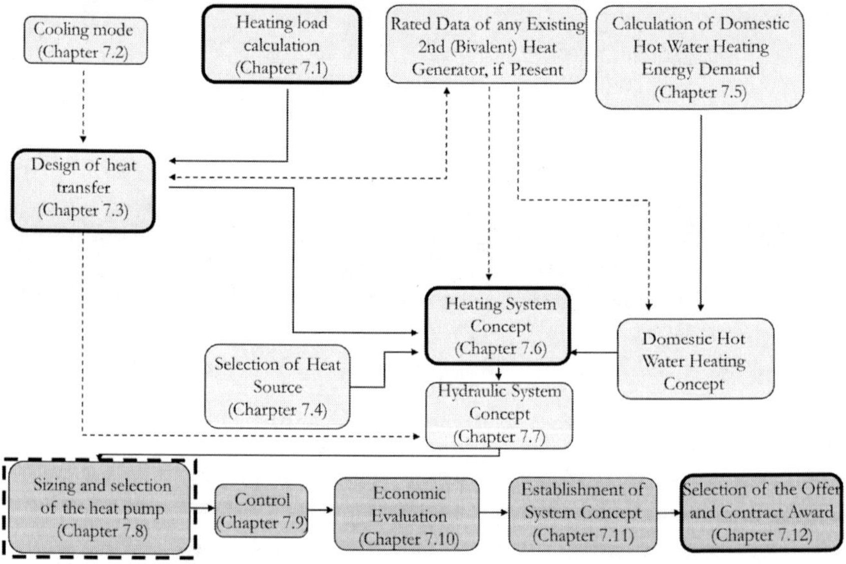

Figure 65: Step-By-Step Guide of a heat pump system, sizing and selection of the heat pump

After the established hydraulic concept (see Chapter 7.7), we now turn to the sizing and prior selection of the heat pump.

For the **sizing of a heat pump**, the design point, depending on the respective mode of operation (see also Chapter 5), is relevant. This clarifies that for the monovalent mode of operation, the monoenergetic mode of operation, or the bivalent mode of operation, different approaches must be taken respectively.

In the monovalent mode of operation, where the entire **heat demand** must be covered by the heat pump at its flow temperature, the design point corresponds to the locally indicated lowest outdoor temperature. For a monoenergetic mode of operation, the design point corresponds to the sizing point. In the bivalent mode of operation, the bivalence point is to be used as the design point.

From the design point chosen according to the mode of operation, the required amount of energy must be provided by the heat pump. To determine the necessary heating output (thermal output) of the heat pump, it is first necessary to determine the individual amounts of heat, which are composed of the heat amount for space heating over 24h at the respective design point, the heat amount for domestic hot water heating over 24h, and the heat amount for other consumers over 24h at the chosen design point. Furthermore, the blocking times of the energy provider (if available) according to the supply contract should also be included in the calculation. However, we want to discuss the four mentioned parameters for determining the required heating output of the heat pump in more detail.

With the calculated heating output (thermal output) of the heat pump, the **selection of a corresponding heat pump** can now be made.

Again, one fundamentally has the option to use specialized software for the sizing of the heat pump. Such software allows the user to simulate various system concepts and operating modes, independent of the heat pump manufacturer. It is possible to input or select all boundary parameters in order to achieve a realistic sizing of the heat pump. Thus, the technical data provided by the heat pump manufacturer for sizing is integrated into the software. By comparing the seasonal performance factors (SPF / HSPF) (see also Chapter 1) and considering costs, it's possible to weigh up the different systems against each other. Caution: for systems operating bivalently, in addition to determining the building's heating load, it's necessary to identify the required flow temperature, which also depends on the heat transfer (see Chapter 7.3).

If it turns out after a few years that the heat pump was sized too small, the worst-case scenario could lead to a loss of comfort. Not least, poorly dimensioned heat pumps have damaged the image of heat pumps in the past. After all, too cool rooms or only lukewarm showers were rightly not what the operator had hoped for from the investment. However, such cases, even nightmarish scenarios, are very rare today; to put it better, they no longer surface. This is largely due to the ever-improving insulation and the rising trend towards radiant heating systems. It is often found, however, that the heat pump is oversized. Heat pump models are always offered in different performance sizes by the respective manufacturer (see Table 6).

For instance, if a required total heating capacity has been calculated and the operator decides on the air-to-water heat pump listed in Table 6, it sometimes happens that planners choose the larger model. The reason for this is the fear mentioned in this approach of a potential undersizing of the heat pump. Now, from a purely numerical perspective, the heat pump is oversized. However, the full heating capacity of the heat pump is only demanded when the minimum standard outdoor temperature is reached. If one re-examines the previous calculation of heating capacity, it takes into account all heated rooms.

Model, series (type)	**8**	**12**	**18**
Rated heating capacity at low-temperature application 95°F (kW)	10	12	16
Rated heating capacity at medium-temperature application 131°F (kW)	9	10	16

Table 6: Example Manufacturer Presentation of an Air/Water Heat Pump in Different Output Classes

While this is correct, practical experience shows that not all rooms are always heated. Often, rooms like the bedroom, a hobby room, or guest room are not heated continuously. This also reduces the actual power demand for the heat pump and can increase the cycling frequency in case of oversizing. As a result, the heat pump is again oversized, leading to a further decrease in the efficiency of the heat pump system, and the compressor is subjected to high stress due to the increased cycling. Moreover, the high switching cycles, caused by the constant restarting of the heat pump, are frowned upon by the grid operators, as the startup currents for the compressor's electric motor are elevated. This can occur solely due to the implementation of unnecessary "fear factors" in the calculation.

However, it should be noted here that inverter-driven heat pumps counteract the issue of oversizing. Unlike fixed-speed heat pumps (also referred to as on/off heat pumps), inverter-driven heat pumps are power-regulated.

If oversizing is done in anticipation of a potential addition that might be made years later, this can also result in significant losses in efficiency. In such a scenario, it's more advisable to install a second heat pump after the expansion. The concept of a heat pump cascade should be considered in this context (see Figure 66). Connecting heat pumps in series can enhance the reliability of the supply through a certain level of redundancy, and by intelligently controlling the heat pumps, power can be drawn more efficiently through a modular mode of operation. As a side note, individual heat pumps are lighter than one large heat pump, greatly simplifying transport and installation.

One can best determine the oversizing by comparing the starts of the heat pump to its operational hours. This becomes even clearer when the compressor's runtimes are divided into load classes. This data can be read from the respective heat pump display.

Believe me, by oversizing your heat pump, you're doing yourself no favors. It's essential to note that, in this statement, the heat pump should be strictly differentiated from the heat source. For heat sources like the ground or water, it doesn't hurt if there's a slight oversizing, for instance, if the ground heat exchangers (see Chapter 4.2.3.2) are designed a bit more generously. Ensure this is again planned by a professional who presents the prepared options with their respective pros and cons, assists you in making the exact decision, and ultimately assumes responsibility with their system design.

Figure 66: Heat pump cascade by WOLF GmbH

When explicitly **selecting the heat pump**, one must now choose the manufacturer and the specific model with its particular performance. At this point, the old saying, "The more choices you have, the harder it is to choose," proves to be true. This can potentially become a complex process for those unfamiliar with the topic. The Appendix C, the heat **pump matrix**, will assist you in choosing the right heat pump.

For research, one can turn to sources such as the internet, trade shows, manufacturers, installers, architects, specialized planners, or magazines. The provider of the particular heat pump isn't always identical to the manufacturer. This shouldn't be seen as a major concern and, in fact, reflects reality. If one is considering purchasing a specific heat pump at a trade show due to a favorable price, a word of caution is in order. At the trade show itself, sales personnel usually offer the heat pumps at a reduced price but may assign the subsequent installation and commissioning to the lowest-bidding installer on a commission basis. While this could work out, it carries a certain "black-box risk," as you can't be entirely sure what to expect. If the installer lacks experience and knowledge, you might be left to fend for yourself and may incur additional costs for expert services to rectify any deficiencies. The initially attractive price becomes a thing of the past, and you find yourself in a dispute before even utilizing the heat pump.

Installers often commit to a particular manufacturer whose models they have been installing for years and in which they have confidence. They then present this model to the building owner to facilitate the decision-making process. There's nothing fundamentally wrong with this approach, as it allows the installer to specialize in a particular brand and become familiar with its system nuances. However, it would be problematic if the installer hands over the entire planning process to the manufacturer and blindly trusts them. It's advisable to pay attention to the mentioned points and critically evaluate any suggestions. Doing so can undoubtedly help determine whether the planning is sound and well-thought-out.

7.9 Control System

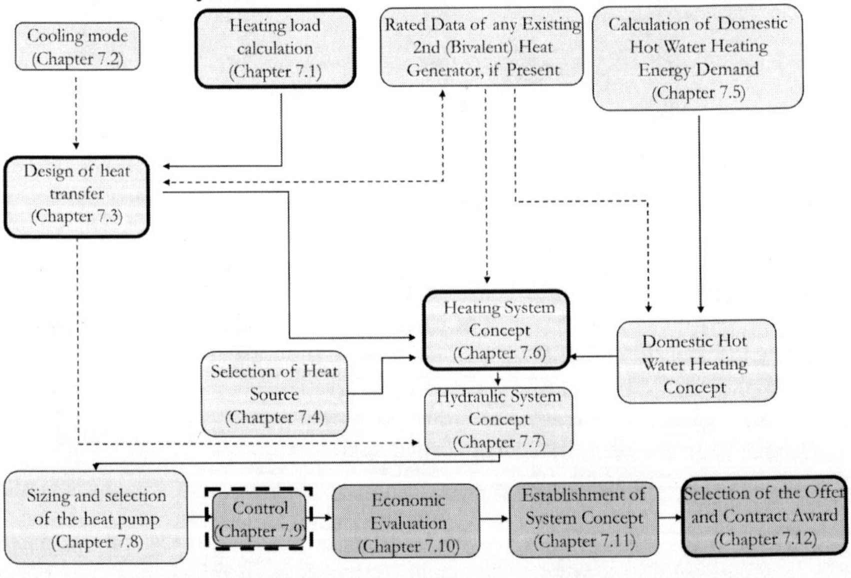

Figure 67: Step-By-Step Guide of a heat pump system, control system

Once the heat pump has been selected and sized (see Chapter 7.8), another crucial aspect in the heat pump planning process is the **control system**. The control system of the heat pump ensures that it provides the right amount of heat, neither too much nor too little. Thus, proper control of the heat pump is essential for energy-efficient and cost-effective operation. The control system covers both the heat source and the heat distribution.

7.9.1 Control of Heat Sources

The **control of the heat sources** can be seen as an integral part of the heat pump system. The control system ensures the **management and monitoring of the refrigeration cycle**. Furthermore, it regulates and monitors the heat source, which involves **controlling components** such as the brine pump, well pump, or fan, depending on the type of heat source. Monitoring the heat source ensures that, for instance, in the case of a brine pump leak, unnecessary brine does not escape due to the associated drop in pressure.

Furthermore, it handles the **control of the heating and cooling circuits**. In the regulation and control of these circuits, multiple heating circuits can usually be managed, with a distinction between mixed and unmixed heating circuits. When regulating the heating circuits, distinctions can be made between outside temperature (also weather-dependent), room temperature, or fixed temperature control. Typically, it involves outside temperature-controlled supply temperature, taking into account room temperature. In weather-dependent control, the heat output of the heat pump adjusts to the outside temperature. The heating curve implemented in the system (see also Chapter 7.3) defines the relationship between outside temperature and supply temperature, determining a corresponding supply temperature for each outside temperature. Thus, it involves variable supply temperature. For instance, if a room temperature of 72°F is desired, the supply temperature is increased, considering the current outside temperature, to reach the desired room temperature. The supply temperature still experiences a necessary increase, not an excessive one, which is more efficient. As building envelope insulation continuously improves, the time lag between room temperature and outside temperature becomes longer. This is often referred to as the room temperature "lagging" behind the outside temperature. Consequently, the heat pump may respond based on the outside temperature even though the room temperature has not yet reached the desired setpoint. The actuator valves of the room controllers are not yet open at this point, causing the heat pump to shut off shortly after starting operation. The "lagging" effect is particularly pronounced with wet-laid floor heating systems due to their thermal mass. Sometimes, heat pump controllers are set to a fixed value. This is common with air-source heat pumps, where the buffer tank is used to defrost the evaporator through a changeover valve. For this purpose, the buffer tank is often set to a relatively high base temperature to ensure sufficient heat supply for the defrosting process at the evaporator. For regulating the cooling circuits, a distinction needs to be made between cooling with and without dew point temperature control.

In the **regulation of domestic hot water heating**, various concepts can be chosen (see Chapter 6).

Finally, the regulation of the heat generators ensures the control of additional heat generators in monoenergetic or bivalent operation mode.

7.9.2 Control during Heat Transfer

In the regulation of heat transfer, it is necessary to differentiate depending on whether it is underfloor heating or radiators.

In the case of **underfloor heating**, which has lower surface temperatures compared to radiators, a **self-regulating effect** occurs. The self-regulating effect directly influences the heat output, which is nearly proportional to the temperature difference between the surface temperature of the heating element and the room temperature. For example, in a well-insulated building, the floor surface temperature for underfloor heating during the heating season averages around 105°F. However, if the room air temperature rises from 100°F to 102°F due to factors like sunlight, the heat output decreases by one-third. Similarly, lowering the room temperature results in an increase in heat output. The self-regulating effect in underfloor heating assists in achieving a comfortable indoor climate from a control perspective.

In the case of **radiators**, room-specific regulation based on individual room temperature is also possible. When underfloor heating is combined with radiators, a separate approach is needed. For example, in a building where only specific auxiliary rooms have radiators and the majority of rooms are equipped with underfloor heating, it may simplify the hydraulic system to supply the radiators with the higher pressure differential from the underfloor heating. This should be taken into account when sizing the valves.

Another method of **heat pump control is through the circulation pump**. The circulation pump keeps the heating circuit in motion, directing the heated water with the supply temperature to the heating surfaces and returning the cooled water with the return temperature to the heat pump. The circulation pump operates through three different control modes. First, the circulation pump switches off when there is no further heating demand. It switches back on when heating demand arises again. If the control is time-dependent, regulation can occur after a predefined time interval. Alternatively, it is possible to control the circulation pump using a timer, which allows the heat pump to be turned off during a predetermined time period. Lastly, the circulation pump switches off when the outside temperature is above the return temperature.

It is also true for the heat pump control, that errors in the settings can significantly reduce the efficiency of the heat pump. If the set temperatures in the buffer tank are too high, this not only leads to poor efficiency but also places additional strain on the heat pump compressor, which, in turn, affects the lifespan of this component. The development and adjustment of a well-coordinated control process should be handled by an experienced institution and only by them. **This means that the responsibility should rest solely with one entity, especially in cases of questions or potential legal disputes.**

Access to the heat pump controller is often possible via the internet using remote maintenance nowadays. For this, an internet connection must also be available in the basement, with a wired internet connection being recommended.

If a building has different users such as offices, workshops, residential areas, and wellness areas, it is advisable to control multiple heating circuits with different temperatures and separate **mixers** over time. However, errors can also occur in regulating system temperatures. It must be ensured that the temperature of the buffer in the heating circuit with the highest designed supply temperature in the system is maintained. However, caution is required, as even a single radiator with an elevated supply temperature can lead to significant efficiency losses. If the heating circuits are activated at different times, differentiation can also be applied to the system temperature.

In general, the principle of following the simplest control concept should be adhered to here, as it is known to work most efficiently in practice. With many heating circuits regulated to different temperature levels and time profiles, there is a risk of inefficiency in the heating system.

7.10 Economic Analysis

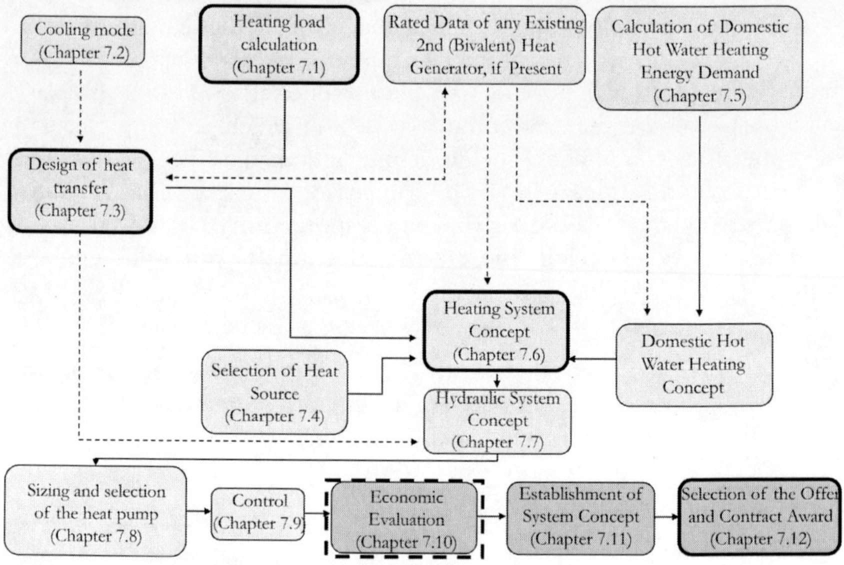

Figure 68: Step-By-Step Guide of a heat pump system, economic analysis

The economic viability of the system is essential. To ensure this, we will now provide a guideline for such an analysis.

While the heat pump tariff offers advantages through discounts, it also comes with some disadvantages. Heat pump operators will incur additional costs due to the extra electricity meter and the ripple control receiver (RCR). Furthermore, the use of time-of-use restrictions necessitates a buffer tank or a larger buffer tank, which, in addition to the limitations imposed by the utility's peak load times, leads to additional expenses.

Generally, the lower the electricity consumption of the heat pump, the less likely a heat pump tariff will be cost-effective. To determine if the heat pump is cost-effective, the initial extra expenses incurred compared to a gas or oil heating system must be recouped within the operating lifespan of the heat pump. **Profitability** is derived from the purchase costs and ongoing expenses.

In the first step, consider the acquisition costs of the complete heat pump system. In the second step, calculate the ongoing expenses (electricity, maintenance, etc.). Once this is determined, an equivalent calculation for the comparison system (e.g., oil or gas heating) is needed. The annual costs of the comparison system (oil, gas, chimney sweep, etc.) must also be listed. It significantly affects the outcome whether, for instance, an existing building's oil heating system needs replacement or can continue to be used for several more years.

Annual costs for electricity, gas, or oil depend on various factors and the current political situation. General forecasts during this time, however, indicate a tendency toward increasing energy prices.

To proceed, let's conduct a comparative analysis using currently prevailing prices as a basis. For this purpose, we assume an operating lifespan of 20 years, which is a conservative estimate since heat pumps are typically used for more than 25 years.

As previously mentioned, the (Heating) Seasonal Performance Factor (SPF / HSPF) of a heat pump describes the quotient (SPF = electrical energy quantity / heating energy quantity) of the resulting annual heating energy quantity (in kWh) to the resulting annual electrical energy quantity (in kWh). SPF is unitless (e.g., SPF (HSPF) = 9.7). But what is the difference compared to the efficiency of a heat pump? The efficiencies of gas or oil heaters are expressed in percentages and roughly correspond to the values listed in Table 13. The efficiency of a heat pump depends on the temperature difference between the energy source and the heat pump's supply temperature. Typically, the efficiency of a heat pump is calculated using the **Carnot formulas**.

7.11 Establishment of System Concept

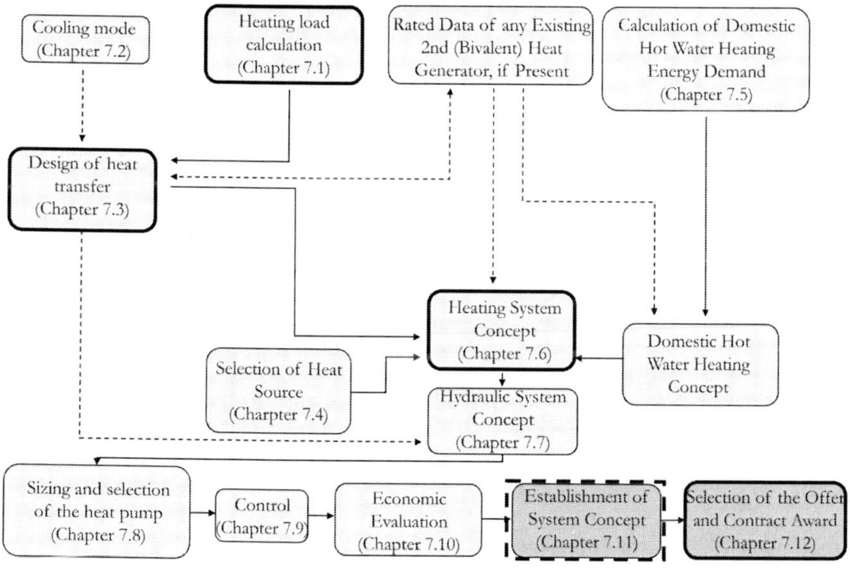

Figure 69: Step-By-Step Guide of a heat pump system, establishment of system control

With the information previously gathered and the various system concepts developed (Chapters 7.1 - 7.9) along with the determination of their respective economic viability (Chapter 7.10), the optimal system concept can now be established. It is possible to make a decision for a future system concept concerning its hydraulic concept (Chapter 7.7), its energy efficiency, and economic viability, among others. These decisions should be documented in writing, including the rationale behind them. This allows you to revisit the decision-making process in the future or evaluate it based on new circumstances. It should be noted that, especially with the widely used air/water heat pump, noise emissions should not be overlooked.

7.12 Selection of the Proposal and Awarding of the Contract

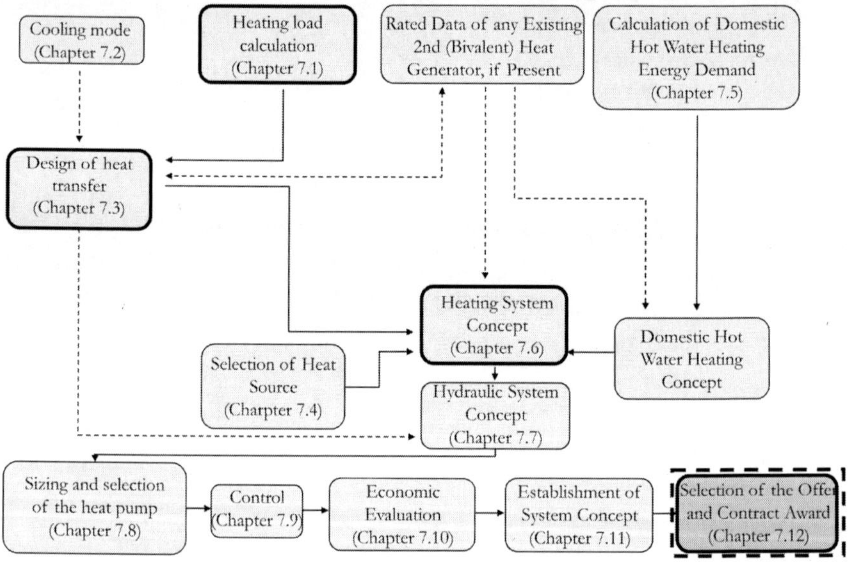

Figure 70: Step-By-Step Guide of a heat pump system, selection of the proposal and awarding of the contract

In this chapter, we will address the process of obtaining and awarding proposals. The determination of the system concept has already been made (Chapter 7.11), and now it is possible to solicit proposals and, after selecting one, proceed with awarding the contract. You might think that this is the relatively simpler part of the planning process, involving just obtaining a few proposals and making a quick rough comparison to choose the most cost-effective one. However, I must disappoint you, as this stage requires the utmost attention. Small errors at this stage can result in significant costs. A heat pump system is a substantial investment and can incur substantial expenses for homeowners. With such prices, mistakes can have a noticeable impact on costs.

For this reason, this chapter will provide you with some advice on obtaining and awarding proposals to help you avoid errors and cost drivers.

- A heat pump should be viewed as a long-term investment, which means that individual components or parts may need to be replaced over time. For this reason, these parts should also be available for the long term, and the question should be asked whether replacement parts will still be available for the chosen heat pump system in over 15 years. Additionally, service and maintenance for the selected heat pump system should also be secured.
- Always inquire about the certification, qualification, and specialization of the provider.
- The individual items should be detailed in the proposal and accepted at a fixed price. This helps avoid any unexpected additional costs.
- As mentioned earlier, the price of the proposal should not be the sole determining factor, and the search for the cheapest offer should not be overly focused. If the heat pump system is generally too expensive, consider choosing a different heat pump system with a different heat source. Also, do not initially include subsidies as a fixed part of the financing, as they are not guaranteed, and the subsidy conditions can change quickly.
- In the proposal, pay attention to price stability, deadlines, guarantees, cash discounts, and warranties.
- Seek a reputable heating contractor (see Figure 51 for areas of expertise) whom you can trust. Perhaps you already have one with whom you have had good experiences for many years. Why not turn to them? Online reviews are often available, but be cautious with anonymous reviews. Even if you suspect that the reviews are "embellished," this should be seen as a warning signal and may lead to the decision not to choose that provider.
- Investigate and, if possible, independently search for references from the provider.
- Consider the use of "natural" refrigerants.
- Obtain multiple quotes.
- For a ground source heat pump system, it is best if the ground loop system and the heat pump come from the same source, as this ensures a single responsible point of contact and eliminates potential

issues with interfaces or the delegation of responsibilities. This applies to both the installation and later maintenance and servicing.

- It's advisable to have the proposals reviewed by a knowledgeable individual (a specialist). This minimizes errors and provides additional security.

Summary:

Planning a heat pump system lays the foundation for your home heating. In this chapter, the individual steps for planning a heat pump system are outlined gradually and supported by small examples. The planning steps for the heat pump system include heating load calculation, potential cooling operation design, heat delivery design, heat source selection, determining the energy requirements for domestic hot water, heating concept, hydraulic concept, dimensioning and heat source selection, control, economic assessment, determination of the system concept, and finally, the choice of proposal and the ultimate awarding of the contract. Please do not assume that you will go through the entire process in one go, step by step, and be finished afterward. Instead, view it as an iterative process in which you can revisit previous planning steps based on new insights or changed opinions to develop the optimal solution for you.

8 FROM INSTALLATION TO COMMISSIONING

In this chapter, the essential aspects of the installation and commissioning of the heat pump system are discussed. Because even if the planning (Chapter 7) was carried out with meticulous care, subsequent aspects such as installation and commissioning can have a significant influence on the efficiency and effectiveness of the heat pump system. Given that this is a multi-layered process, this chapter is further broken down into individual primary stages of installation and commissioning in the form of sub-chapters, ensuring that you always have a clear overview.

8.1 Filling, Venting and Pressure Test

Regarding filling, venting, and the pressure test, the requirements vary from manufacturer to manufacturer. It is imperative to adhere to these requirements. Otherwise, it might happen that the technician assigned for the commissioning of the heat pump system cannot perform his tasks upon arrival.

Filling the heat pump system should not be underestimated. If the local water has too much hardness, it's often sufficient for the installer to use portable softening cartridges. It is essential that you, as the operator, are informed about this and that during subsequent fillings or top-ups, the water quality specified by the heat pump manufacturer is maintained. A better solution is to have a permanently installed water softening system based on a water analysis to counteract the excessive hardness of the water.

Following this, the heat pump system requires venting and a pressure test by the installer.

8.2 Sizing and Installation of Pumps, Pipelines, Fittings and Safety Devices

The heating operation relies on the correct sizing of the components surrounding the heat pump, such as pumps, pipelines, fittings, and safety devices. For sizing the pipelines, manufacturers' planning manuals can be utilized. These manuals contain tables with minimum flow diameters depending on the heat pump performance. This ensures the risk of undersizing is eliminated, which would result in non-compliance with minimum flow rates, reduced heat transfer, incorrect settings of the circulation pump (see Figure 71), and efficiency losses.

If significant temperature differentials (temperature differences) between the supply and return lines occur, it's often due to undersized pipelines. The high temperature differential is then commonly countered with a circulation pump set to its maximum delivery height or an oversized circulation pump.

Figure 71: Exemplary circulation pump

Even without a conducted hydraulic balancing (see Chapter 8.5), the circulation pump serves as a compensation, which can result in reduced efficiency and increased electricity consumption. This increased electricity consumption might not be noticeable, especially if the circulation pump is not connected to the special power circuit of the heat pump.

A dirt and sludge separator should also be installed. Precipitations can form in the pipes, storage tanks, and heating surfaces, leading to deposits at elevated temperatures, which impair heat transfer. Blockages in the pipelines or contamination of the heat pump's evaporators or condensers can also occur, resulting in efficiency losses. Such incidents are prevented by dirt and sludge separators.

8.3 Insulation

An efficient heat pump system requires good insulation. This includes proper insulation (see Figure 72) of pipelines, pumps, and fittings, according to the regulations.

Figure 72: Exemplary insulation of the heating pipes

All too often, in existing heating pipes, one encounters insufficient or even missing insulation. Especially in areas that aren't readily visible, inadequate insulation is not uncommon. In some areas of the floor, even with the heating turned off and during summer temperatures, warming can occur due to insufficient or missing insulation. The consequence of this is energy loss that's absorbed by the building, leading to an inadequately warm water temperature at the tap point of the drinking water line in a short amount of time.

If condensation can form on the pipes, they must be insulated in a diffusion-tight manner. For cold pipes, it's advisable to install them using the shortest possible routes. This should especially be considered for the brine lines of ground-source heat pumps.

8.4 Commissioning

When commissioning, it is crucial to use the commissioning protocol provided by the heat pump manufacturer. In many aspects, these protocols are similar across different manufacturers. However, always ensure to use the manufacturer-specific commissioning protocol. This protocol helps verify whether all the listed points related to the manufacturer's warranty have been addressed, ensuring that the heat pump was planned and constructed according to the manufacturer's guidelines. This protocol assures the manufacturer that potential errors related to adhering to planning and installation instructions have been avoided, thus eliminating future damages to the heat pump.

Maintaining this commissioning protocol also ensures the efficient operation of the heat pump.

Several checkpoints must be considered and listed in the commissioning protocol when commissioning heat pumps:

- Is the rotation direction of the compressor correct?
- Has the electrical connection been implemented according to the electrical plan?
- Was the system installed following the manufacturer's hydraulic plan?
- Are the minimum flow rates compliant with the guidelines?
- Have the transport safety devices been removed?
- Was the system filled with the water specified by the manufacturer?
- Were dirt traps installed before the heat exchanger?

Different approaches can be taken for commissioning. In most cases, the heat pump's commissioning is carried out by the manufacturer's customer service. Another option is commissioning by specific installation companies with their service technicians. If this method is chosen, always consult with the heat pump manufacturer to ensure that certain warranty services aren't exclusively valid when the manufacturer handles the commissioning. Another option is a certification issued by the manufacturer to an installation company, meaning that the service technicians have been trained for that specific heat pump model.

Therefore, when commissioning the heat pump, pay attention to the appropriate qualifications of the person commissioning it and ensure that the commissioning protocol is fully and correctly filled out. Finally, keep the commissioning protocol in a safe place.

8.5 The Hydraulic Balancing

Efficient heating operation is critically reliant on hydraulic balancing. However, in the past, a striking number of heat pump operators neglected this step. Estimates suggest that hydraulic balancing was not performed in about 80% of all heating systems. Indicators of this oversight include circulation pumps set to maximum delivery heights, identically set regulating valves, elevated return temperatures, or over- or undersupplied heating surfaces. But what exactly is hydraulic balancing? The VDI 2073 Part 2 provides a detailed description of hydraulic balancing, ensuring that the existing heating surfaces or radiators in the heating system are supplied uniformly. In a hydraulic system consisting, for example, of various radiators, a consistent flow rate should be present at each radiator. This means that radiators situated farther from the heating system tend to be undersupplied when, for instance, almost all radiators in the system are opened during morning heating. The undersupply of distant radiators can be exacerbated by unfavorable behavior of the thermostat valves. To counteract this, the flow of closer radiators is deliberately reduced, ensuring a balance is achieved and uniform supply to the entire system is consistently maintained.

Yet, revisiting the common practice of setting the circulation pump to its maximum delivery height, attempting to replace hydraulic balancing by trying to raise the flow temperature and thus compensate for discrepancies in the heating circuit distribution, the intended effect is an oversupply of heating surfaces with low-pressure loss, increased power consumption of the pumps, and reduced coefficient of performance. Increasing a circulation pump's delivery height to raise the volume flow, for instance, doubling it, results in an approximate 12% increase in heat output. This underscores that only a professionally conducted hydraulic balance ensures the desired uniform heat supply in the heating system.

One might then easily question why one would meticulously plan a heat pump system, incurring considerable costs, if such vital steps as hydraulic balancing are overlooked during installation. A frequent reason for this is a lack of professional knowledge or simply opting for cheaper work. If the heating technician omits adjustable thermostat valves, heating circuit distributors with flow meters, differential pressure valves, or return screw connections, this can significantly affect the offer (for related fields, see Figure 51 in Chapter 7).

Performing hydraulic balancing in a new building is typically straightforward since calculations for heating surfaces and the corresponding volume flow requirement are available. Usually, volume flow regulation is also considered.

This is not the case with existing systems. For example, if one wants to perform hydraulic balancing of an underfloor heating system, old implementation documents must be obtained, provided they still exist. From these, information about installed pipelines or installation intervals can be extracted. If there are no implementation documents, infrared cameras or infrared thermometers can be used to determine the respective installation interval, and from this, the existing heating circuit length can be calculated.

Through hydraulic balancing, the heating output is fully transmitted, the rooms are heated evenly, room temperature fluctuations are reduced, noise disturbances from fittings and other components are avoided, and the power consumption of the circulation pump is decreased due to reduced power demand.

Also, hydraulic shortcuts should be avoided. In the case of hydraulic shortcuts, there is a very small temperature difference (temperature difference) between the flow and return. For surface heating systems, temperature differences between 5 and 10 degrees Fahrenheit are used as a guideline, while for radiators, values between 15 and 20 degrees Fahrenheit are appropriate.

Please also note the requirement for hydraulic balancing when applying for funding.

8.6 Instruction and Hardcover to the Operator

After the installation has been fully completed, there is a need for instruction and handover to the operator of the heat pump system. I recommend that you familiarize yourself with the heat pump's manual in advance, so you're not going into the instruction completely uninformed. This has the advantage of allowing you to ask specific questions, which might not arise if you were completely unfamiliar with the system. Finally, after the handover has taken place, it should be documented in writing. The document should include a confirmation of the completion of the agreed-upon services. If any defects have become apparent during the handover or even earlier, they should also be documented and counter-signed in the handover protocol.

Summary:

After filling, the system must be properly vented. The piping must be correctly dimensioned, and dirt and sludge separators must always be installed in the system. The system must be adequately insulated to prevent condensation. An operation protocol from the heat pump manufacturer should be maintained and subsequently stored securely. The person initiating the operation should have sufficient qualifications. To ensure efficient heating operation, the **hydraulic balancing** is indispensable. Familiarize yourself with the heat pump's manual before the instruction session.

9 MAINTENANCE, INSPECTION AND SERVICING

To maintain a functional and energy-efficient heat pump system while adhering to environmental protection standards for an extended period, regular **inspection** and **maintenance** are crucial. This ensures the proper operation of components with limited lifespans and preserves the efficiency of the heat pump system.

It is essential that the personnel responsible for **maintenance** and **servicing possess** the necessary expertise and experience. The system should be operated correctly by the owner, and components that significantly affect the system's efficiency should be regularly maintained and, if necessary, repaired.

For modern heating systems with integrated heat pumps, inspection and maintenance can be combined, divided into two areas: the **general part** for the entire heating system and the **specific part** focusing solely on the heat pump. Depending on the findings, maintenance may be required.

For the **general part**, an inspection of expansion tanks, safety valves, circulation pumps, component fastenings, and a check for damage and contamination of the heating system are performed. Additionally, a cleaning of the heating circuit filters is carried out. When reading the fault memory and operational data, indications of potential issues may arise, with specific controller settings being crucial for optimizing the heat pump's efficiency. Special attention should be given to possible oil traces around the heating system, as they can be an indicator of refrigerant leaks. A **leak test of the refrigeration circuit** is recommended at legally required intervals if the heat pump is subject to the Chemical Climate Protection Ordinance due to its refrigerant filling quantity. Regular leak tests are intended to prevent unintentional releases of large amounts of greenhouse gases. The heat pump's refrigeration circuit is exposed to continuous stress due to the compressor's vibrations, and refrigerant leaks can indeed occur. The heat pump may not always detect refrigerant leaks as malfunctions, and the system's efficiency can suffer as a result. In some cases, there may be issues such as incorrect shutdowns in the system. Furthermore, the operator is responsible for complying with leak test intervals in the system and must ensure that a certified company, which documents the results of the leak test in accordance with regulations, performs the leak test. A maintenance contract, for example, is suitable for this purpose. More detailed information on the required certification will be provided at the end of this chapter to help operators avoid unnecessary errors.

Specific **maintenance** and **inspection** must be carried out according to the respective type of heat pump:

- For water-water heat pumps, if there are muddy deposits in the filters, it indicates a changing water quality, which can often lead to impaired heat pump function. In this case, upgrading with a system separation is recommended.
- The ground-source heat pump requires an inspection of the antifreeze protection and the possibly installed ground pressure switch.
- With an air-water heat pump, cleaning of the evaporator and condensate discharge is necessary.

But who should be responsible for inspection, maintenance, and upkeep? Clearly, the person must be certified for these tasks. If the heat pump system is planned by a company, they are responsible for inspection, maintenance, and upkeep at least during the warranty period. In practice, there are quite different constellations in the collaboration between heating engineers and refrigeration technicians. Heat pumps are typically manufactured and distributed by heating industry manufacturers. They deploy service technicians, trained in the entire heating sector, for inspection, maintenance, and upkeep, depending on the sales area. These service technicians usually have extensive experience but may not have the same depth of knowledge as refrigeration technicians when it comes to problems in the refrigerant circuit. Alternatively, manufacturers may train their service personnel and deploy them across the board. Ideally, the service technician for the heat pump system is an experienced refrigeration technician with qualifications and experience as a heating engineer (see Figure 51 in Chapter 7 for areas of expertise). An explicit separation of refrigeration technician and heating engineer in the heat pump sector is usually suboptimal since it often leads to interfaces and potentially longer assignments of responsibilities in the event of a fault analysis. It is also rare for the heating engineer and refrigeration technician to conduct a fault analysis together on-site, further exacerbating the interface problem described.

Summary:

Both maintenance and repair are very important tasks that require the executing trade to demonstrate the appropriate expertise and experience. In modern heating systems with integrated heat pumps, inspection and maintenance can be combined, with inspection and maintenance being divided into two areas: the general part for the entire heating system and the specific part that only concerns the heat pump. Depending on the findings, maintenance may be required.

PART III

«The crucial of knowledge is, that one internalizes and applies it.»

-Konfuzius-

10 COMBINATION OF ALTERNATIVE ENERGY SYSTEM

The synergy of the three components – **photovoltaic system**, **heat pump**, and **storage** – will become increasingly attractive in the future for three main reasons. In the long run, the procurement costs of both the photovoltaic system and the battery storage (also called electricity storage, see Figure 73) are declining, making the acquisition of these two components more favorable. However, in Germany, the feed-in tariff for photovoltaic systems in private households, which is set by the EEG, is steadily decreasing, while the costs to obtain electricity from the energy supplier are rising. This leads to the conclusion that it's most economical to use self-generated electricity for one's own needs. This is precisely where the third component, the heat pump, comes into play. By utilizing the three components of photovoltaic systems, battery storage, and heat pumps, a building's self-sufficiency rate can be increased to up to 65%.

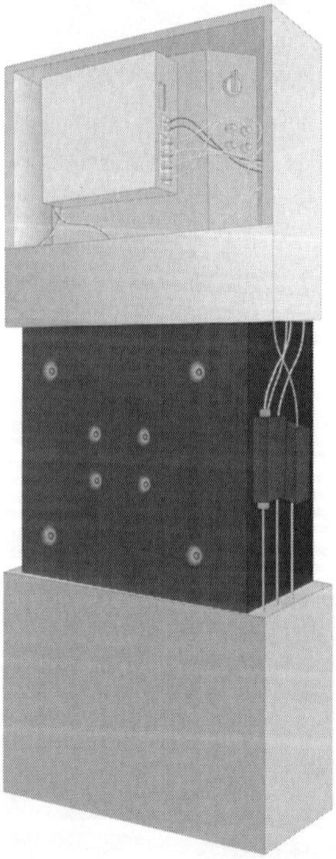

Figure 73: Exemplary electricity storage

Figure 74 illustrates an exemplary interaction of the three components.

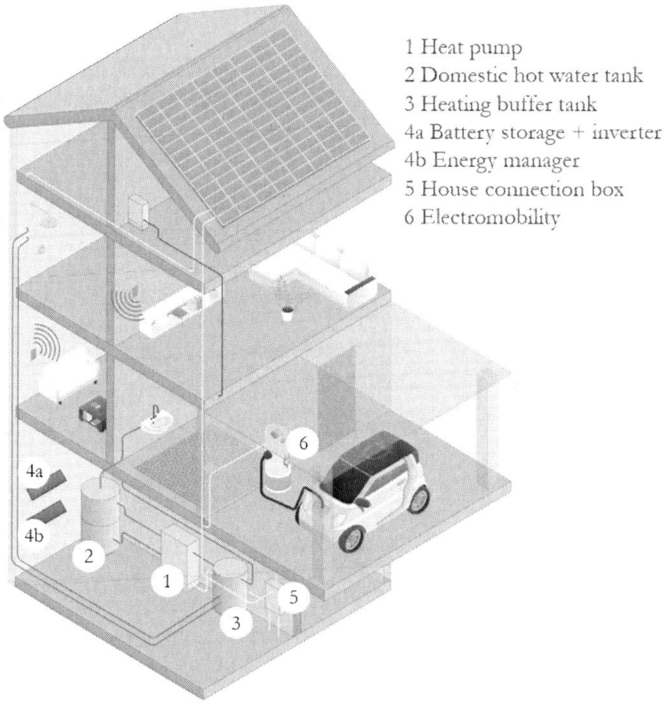

Figure 74: Interaction of the three components: photovoltaic system, heat pump, and storage

The photovoltaic system generates the electricity needed to operate the heat pump as well as for other consumers in the house. Excess electricity produced by the photovoltaic system is used to charge the battery storage. Alternatively or additionally, a thermal storage could be implemented here, which, in combination with the battery storage, converts excess electricity into thermal energy using the heat pump and stores it.

The stored electrical energy can be used directly from the battery storage on subsequent days. Similarly, the stored thermal energy is used for heating purposes or for hot water supply from the thermal storage. In general, it can be said that, based on the current state of technology, thermal storage has a greater storage potential than battery storage.

Innovative inverter heat pumps can operate at variable speeds, offering the significant advantage that reducing the speed also reduces the electrical power consumption.

An electric vehicle could also be charged using electricity from the photovoltaic system.

To ensure the coordination of the mentioned components and to produce an optimum result, so-called energy managers are required. Through these energy managers, "peak shaving" is ensured, which in turn provides financial incentives for battery storage systems. **Peak shaving**, also known as load peak reduction, is a method for load control. Through this, peak loads are smoothed or stabilized by load shifting, load shedding, and energy storage systems. The energy manager thus controls the generation, storage, and consumption of the entire system. An important aspect of the energy manager is that it makes forecasts and organizes the workflows between the components based on these predictions. This will be further discussed with an example of a daily cycle in the fall.

The thermal storage was used to operate the heating system at **night**. With the first rays of the sun in the **early morning** (see Figure 75), consumers are powered by the photovoltaic electricity. Here, the battery storage supports the operation of the heat pump. And now, the aforementioned solar power forecast in the energy manager comes into play. After the morning hot water demand in the household (e.g., showering), it determines whether it is worth starting to reheat the hot water now or waiting until sufficient electricity can be fed in from the photovoltaic system (e.g., at noon).

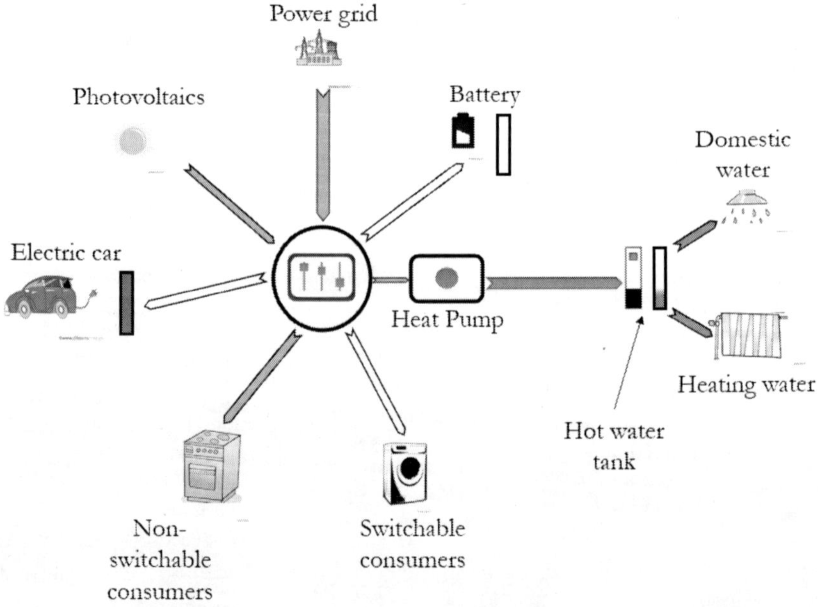

Figure 75: Energy manager display: Early morning

In the **morning** (see Figure 76), the battery storage is charged by the electricity from the photovoltaic system. At the same time, the energy manager reports no need for room heating, so the heat pump is not in operation. By **early afternoon**, the battery storage is fully charged. The energy manager now forecasts that there will be an increased demand for heating energy in the coming days. For this reason, the heat pump is activated based on the electrical energy from solar radiation, and the thermal storage is fully charged.

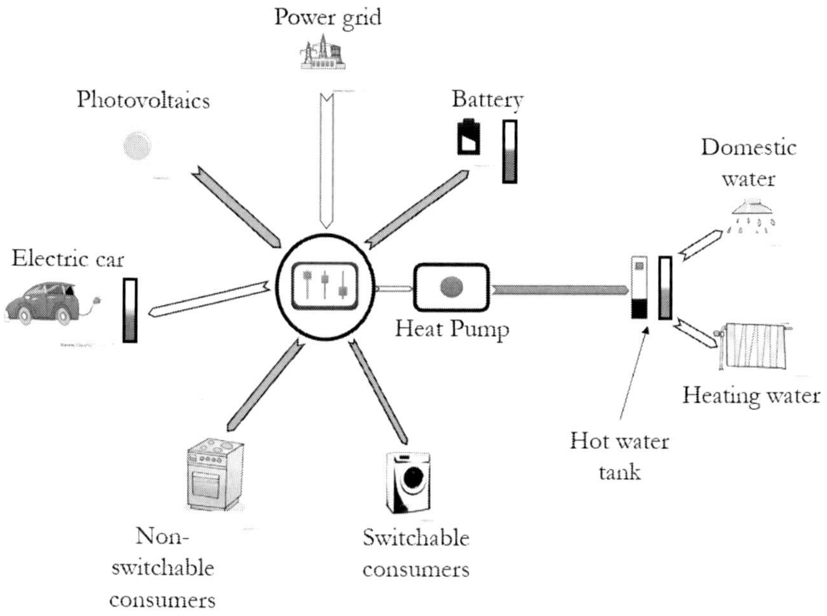

Figure 76: Energy manager display: Morning

In the **afternoon** (see Figure 77), the electric car returns to the garage and is charged using the electricity from the photovoltaic system, while at the same time, the electric energy from the photovoltaic system is used to operate the heat pump and charge the thermal storage. Should there be minor fluctuations in the photovoltaic electricity, the battery will compensate for these. At dusk, in the early evening, the electricity consumption in the house increases and is covered by the battery storage.

Since the heating in the **evening** (see Figure 78) also indicates a need, this demand is met from the thermal storage, preventing the heat pump from operating and thus not requiring electricity. Once the thermal storage is depleted, the heat pump can again be operated using the remaining electric energy from the battery storage until sunrise.

In the given exemplary daily routine, the energy manager operates according to a clear hierarchy of consumption.

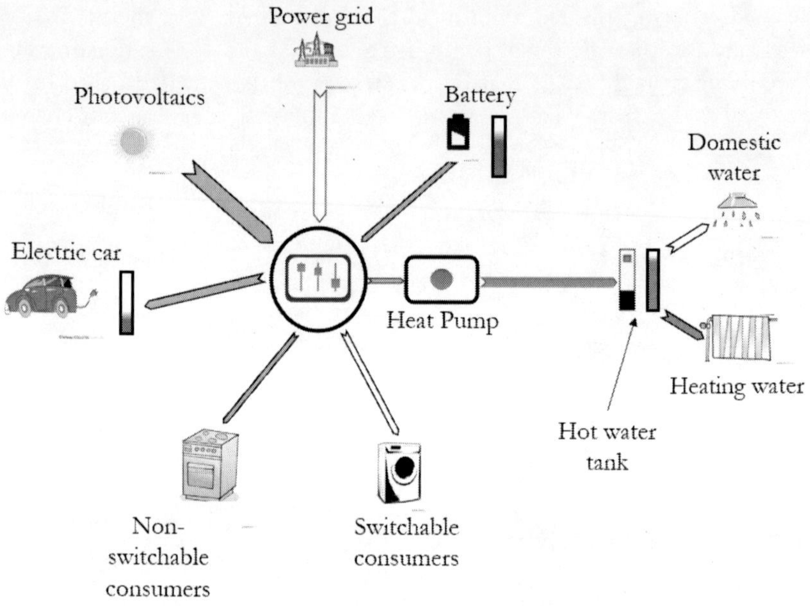

Figure 77: Energy manager display: Afternoon

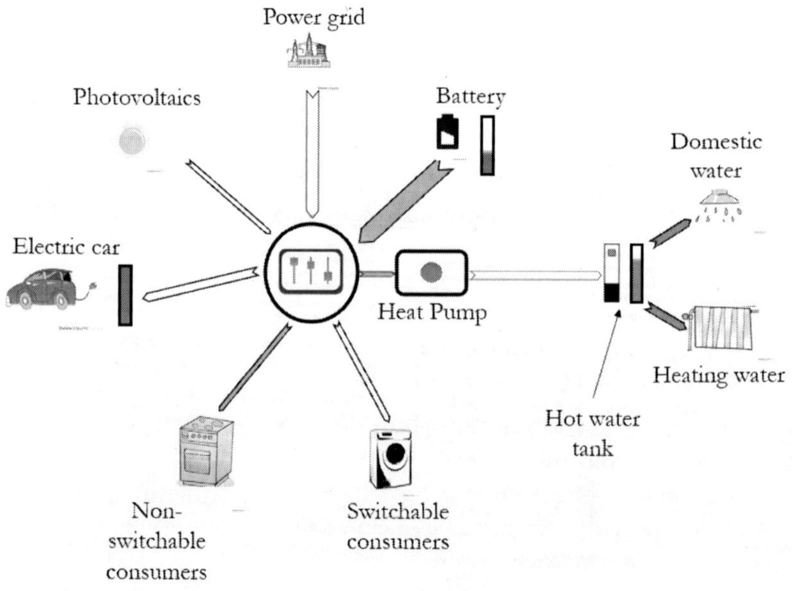

Figure 78: Energy manager display: Evening

For example, if certain minimum temperature values are breached, the heat pump is activated to ensure that the resident always achieves their set comfort temperatures.

Particularly on cold winter days, there might be situations where electricity from the grid needs to be supplemented. But such scenarios should be minimized in the individual design of the components to rare statistical outliers. In this context, the optimal programming of the energy manager, focusing on energy savings, plays a crucial role and will continually evolve in the future.

In the coming years, the energy manager, using locally recorded and meteorological data, will be able to optimize the interplay of the components in each house to achieve minimal energy consumption. This requires a detailed data analysis, such as household shower habits. All this data will be evaluated in an encrypted form and will be subject to strict confidentiality requirements. The future holds limitless possibilities for such developments. Figure 79 provides an example.

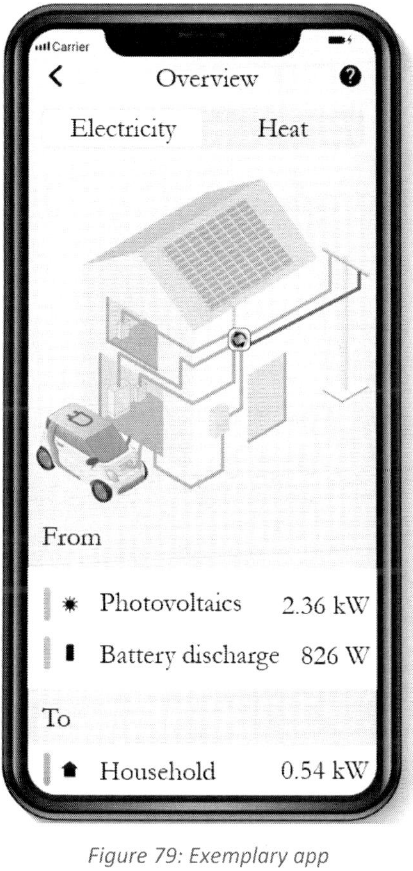

Figure 79: Exemplary app

Summary:

The combination of three components - **photovoltaic system**, **heat pump**, and **storage** - is becoming increasingly attractive in the future, primarily for three reasons. Generally, over the long term, the costs for acquiring a photovoltaic system and battery storage are decreasing, promoting the procurement of both components. With such a system, when paired with an energy manager (electrical unit), you have a tremendous potential for energy savings.

11 TROUBLESHOOTING AND OPTIMIZATION

11.1 Troubleshooting

Errors in heat pumps are distributed across three areas: heat source system (HSS), heat pump (HP), and heat use system (HUS) (see also Chapter 2.1).

Errors can arise **from planning** (see also Chapter 7), **installation** (see also Chapter 8), and **operation** of the heat pump system. We will delve deeper into these three categories.

11.1.1 Errors in Heat Pump System Planning

- Insufficient design of the heat source
- Planning mistakes in hot water generation
- Heating surfaces are wrongly laid out by incorrectly setting the flow temperature or misinterpreting or neglecting thermal resistance in underfloor heating
- No or faulty heat load calculation was carried out, leading to the wrong dimensioning of the heat pump
- In the water/water heat pump system, the pump for the flow rate on the source side was oversized
- Connection line for the brine/water heat pump system was undersized
- Heat source for the brine/water heat pump system was wrongly dimensioned
- Brine circulation pump of the brine/water heat pump system was incorrectly designed
- The noises of the air/water heat pump system are too high, or the position or the potential flow rate of the lines was not planned correctly
- For systems in bivalent operation, only the building heating load was used for dimensioning the heat pump, and the required flow temperature was not determined

- In systems in bivalent operation, an unfavorable hydraulic or control concept was used
- Sufficiently qualified personnel were not involved in the planning (includes previous points)

11.1.2 Errors in Heat Pump System Installation

- There is no hydraulic plan, or it is inadequately prepared
- Installed components in the heat pump system are not coordinated with each other, or their compatibility is associated with complications
- The pressure in the heating system is set too high, or the expansion tank is undersized
- No hydraulic balancing was performed, or it was done inadequately
- No proper instruction and handover were conducted for the operator
- In the brine circuit of the brine/water heat pump system, there are air or impurities present
- The condensate drain of the air/water heat pump system was not implemented frost-proof
- The insulation is insufficient. In addition to the pipes, pumps and flanges also need to be insulated.

11.1.3 Errors in Heat Pump System Operation

- The heating curve is set incorrectly
- The external temperature cut-off limit is set too high on the heat pump controller
- The holiday mode of the heat pump system is not being utilized
- Night setback is activated. Frequent operational interruptions and the associated lowering of room temperature lead to higher consumption
- The hot water preparation takes place during the day when the outdoor temperature is at its lowest for air heat pumps
- Hot water preparation does not occur during off-peak hours when

operating with special electricity rates

- No switching between winter and summer operation
- No data collection and analysis from previous years
- The external temperature cut-off limit is set too high
- Wall or underfloor heating is inefficiently covered
- No proper instruction and handover were conducted for the operator
- In the water/water heat pump system, too high a flow rate is set on the source side
- The brine concentration of the brine/water heat pump system is too high

11.1.4 Possible Malfunctions and Causes in Heat Pump Systems

Table 7 succinctly lists the possible malfunctions and causes in heat pump systems.

High pressure	Low pressure	Lubrication	HP-Control
Heat transfer return temp. too high	Insufficient refrigerant	Oil deficiency	Power outage
Condensation temp. too high	Inlet temp. too low	Oil pressure	Sensor breakage
Regulation	Medium temperature too low	Oil heating OFF	Main switch OFF
Contamination	Contamination	Oil temperature too high	Error SPS
Mass flow rate too small	Regulation	-	Improper operation
Water filter contaminated	Additive control function	-	-
Too much refrigerant	Mass flow rate too small	-	-

Table 7: Malfunctions and Causes of Malfunctions in Heat Pump Systems

11.2 Optimization

By far, the most significant potential for optimization in existing heat pump systems lies in the **heating curve** and the settings of the **heat pump controller**.

For an optimal **setting of the heating curve**, the following step-by-step procedure should be applied:

First, determine the desired minimum temperature for each room. Then, determine which heating surface in which room requires the maximum flow temperature. To do this, the thermostats in all rooms are fully opened. The warmest room in the building is designated as the reference room, which is usually the living room or bathroom. The flow temperature is then progressively reduced until the desired minimum temperature is just about reached in the reference room. With this approach, it is useful to approach the desired temperature from a cooler starting point, gradually warming up to the reference room temperature that just achieves the desired level. If you have underfloor heating, due to its inertia, this process may take several days or even weeks. Even with radiators, some time should be allowed, as this entire procedure must be repeated multiple times at different outdoor temperatures to correctly determine the inclination (also called the slope) of the heating curve (e.g., -14°F, -23°F, 32°F, 73°F, and 82°F).

The more heat output that the heating surfaces can provide and the better the thermal insulation of the building, the flatter the heating curve can be set.

As a basic rule, it can always be assumed that an increase in the flow temperature by 1 °F causes an increased consumption of 2.5%.

With the heat pump controller, there is a whole range of setting options that differ from the standard factory settings. Areas to optimize include:

- The bivalent point and the secondary heat generator
- Circulation pump settings
- The vacation program
- The outdoor temperature cut-off limit
- Switching between summer and winter operation
- The hot water temperature
- Operating times for heating and hot water

However, there are even more optimization opportunities, many of which are reversals of the aforementioned errors:

- Monthly meter readings should always be recorded
- The operating mode should be switched between winter and summer operation
- The vacation program should be used during longer absences
- For a brine/water heat pump system, check the brine concentration at least every two years, and never top off missing brine with water
- In a brine/water heat pump system, temperature measurements of the brine inlet and outlet temperatures give insights into the heat source
- The circulation pump should only be active when hot water is needed
- The outdoor temperature cut-off limit should be set as low as possible. It's best to approach the possible limit from the bottom
- A recurring check of the room temperature should ensure that it truly meets individual needs. Otherwise, the room temperature should be adjusted downwards.
- Check to ensure that the heating output has not been reduced by coverings (see also Chapter 7.3)

CONCLUSION

We have now reached the end of this book on heat pumps and heat pump systems.

Throughout this book, we have been provided with fundamental knowledge about heat pump systems. We are familiar with various energy-related terminology, understand the structure and function of different heat pumps, can differentiate between various application areas, and have learned about the heat sources used in heat pump applications. Furthermore, we have explored the different operating modes of heat pumps and are now aware of the various types of hot water storage tanks.

In addition, we have been given a step-by-step guide for planning a heat pump system that we can refer to repeatedly. This planning process can be used as a blueprint for our individual planning needs.

We have also gained deeper knowledge and insights into potential errors and optimization possibilities.

However, what we have learned about heat pumps and heat pump systems in this book represents only a fraction of the extensive knowledge in this field, as outlined in the introduction. Our goal of providing you with the tools for planning and operating a heat pump system, in addition to foundational knowledge, has been achieved.

For any comments or questions, my email address is available in the imprint. Please feel free to reach out. The energy transition is unstoppable and well underway, compelling us to stay up to date. You now have the opportunity to put what you've learned into practice because, as the German philosopher Immanuel Kant noted, "Theory without practice is empty, practice without theory is blind."

Figure 80: Eureka-Luft (EU-L) series by LAMBDA Wärmepumpen GmbH

APPENDIX- BONUS MATERIAL

A. STEP-BY-STEP GUIDE

Here is the diagram that will guide you in planning your heat pump system. In Chapter 7, you will find a detailed explanation of the Step-By-Step Guide.

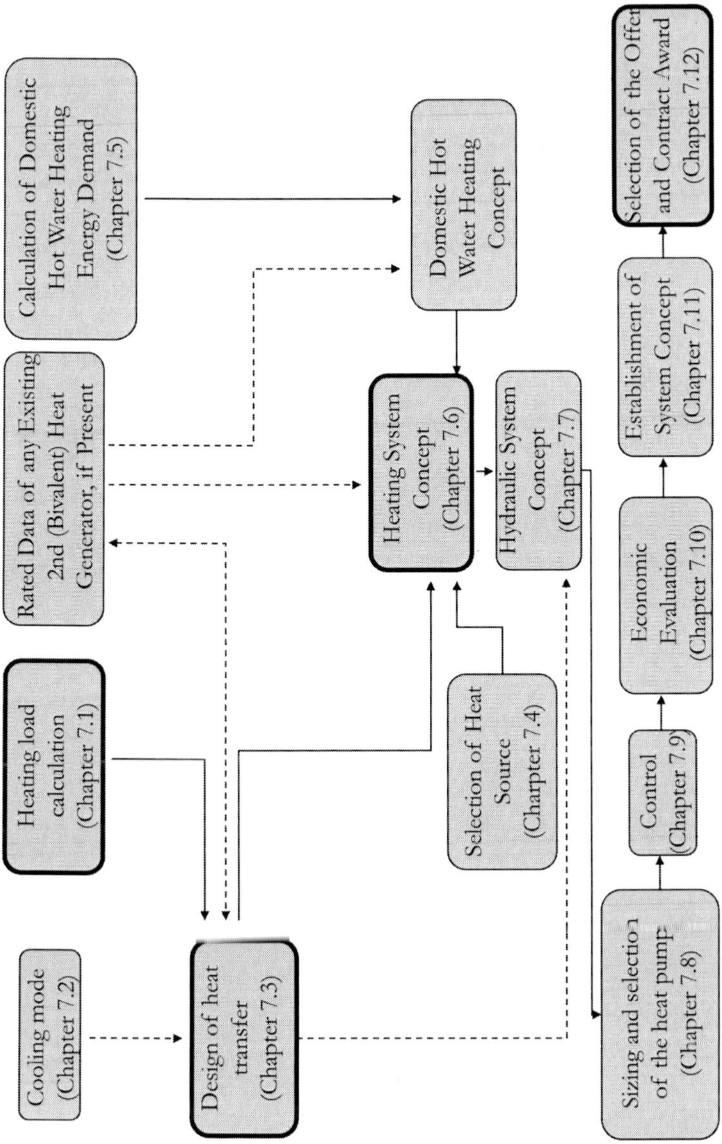

Figure 81: Step-By-Step Guide of a heat pump system, overall view, Appendix A

B. PLANNING CHECKLIST

The checklist for planning, as presented here, is explained in Chapter 7.6 and forms the concept for your heat pump system. This will enable you to use it for your own individual planning.

General	
Contact person	
Addresses, etc.	
Location of construction project	
Standard design temperature / climate zone [°F]	

Building	
Year of construction	
Usable floor area [m2]	
Number of functional units	

Heating load	
Standard heating load calculated	
Determined from consumption data	
Planned renovation of the building envelope	
Planned remodeling/ extensions/ attic conversion	

Type of use	
Residential or residential-like apartment	
Other kind of use	
Hotel/accommodation facility	
Daycare center	
Sports facility	
Medical practice or similar	
Sales rooms	
Other	

Property	
Size/Topography	
Available area	
Soil condition	
Accessibility	
Slope	
Planned/existing embank ments	
Retaining walls	
Distance to neighboring buildings	
Land use	
Tree stock	
Water bodies	
Architectural requirements	
Planned development	

Type of heat source	
Air - Indoor installation	
Installation room	
Air routing, existing and planned openings	
Air - Outdoor installation	
Distance to neighboring buildings (noise)	
Connection to the boiler room	
Condensate drainage	
Ground - Earth heat collec-tor/Earth heat basket, etc.	
Available area for the col lector	
Connection to the boiler room	
Soil condition	
Approval required?	
Ground - Earth heat probe	
Accessibility for special drilling equipment	
Minimum distances to buildings, property bounda ries, other probes	

Connection to the boiler room	
Approval required?	
Groundwater	-
Water analysis	
Pumping capacity/Pumping test	
Connection to the boiler room	
Approval required?	

Heat generator	
Existing heat generator: dismantling/continued use	
Planned additional heat generator	
Operating mode monovalent/monoenergetic/bivalent alternative/parallel	
Installation room for heat pump and accessories	
Function	
Heating	
Heating and cooling	
Domestic hot water	
Swimming pool	
Other	

Heat distribution	
Continued use of existing components	
Necessary measures on the system hydraulics and control technology	

Type of heat transfer	
Type of heating surface	
Function heating or heating and cooling	

Domestic hot water heating	
Reuse of existing components	
Operating mode	

Building cooling	
Operating mode: Tempe-rate/Cooling	
Type of distribution system (surface cooling/air handling)	
System temperatures	
Building cooling load	

Provision and energy supply	
Available tariff models	
Existing infrastructure, e.g., photovoltaics, electric storage heating	
Utility company blackout duration	

Sonstige Nutzeranforderungen	
Future usage profile	
Economic specifications (incorporation of solar thermal, biomass, etc.)	
Other requirements, e.g., comfort-related (required hot water temperature, heating reserve according to DIN EN 12831-1, etc.)	

C. HEAT PUMP MATRIX

Heat Source	Design	Advantages	Disadvantages
Water	Ground source	- Independent of surface temperature fluctuations (weather) - Requires less space - Can be built over - Highest efficiency - Nearly silent	- Very expensive - Approval required - Not always possible
Ground	Ground source	- Independent of surface temperature fluctuations (weather) - Requires little space - Can be covered or built over - Virtually silent	- Expensive - Requires approval
Ground	Ground collectors, Geothermal baskets, trench collector, Ring trench collec-tor,	- More cost-effective than ground probes - Typically no permit required, or a straightforward permit if needed - Virtually silent	- Somewhat dependent on surface temperature fluctuations (weather) - Large space requirement (land area) - Must not be built over
Air	Air-to-water heat pump in compact design, outdoor in-stallation	- No noise level inside the building - Less space required in the building	- Increased noise level outside the building possible - Foundation and free condensate drainage required - Wall penetration required - Heat losses through the connecting line
Air	Air-to-water heat pump in split design, compressor and evaporator outdoor installation	- No noise level inside your own building - No risk of freezing - Wall and floor installation possible	- Increased noise level outside the building due to compressor possible - Condensate drainage required
Air	Air-to-water heat pump in split design, compressor indoor installation, evapora-tor outdoor installa-tion	- Low noise level due to fans and evaporator, outside - Unproblematic condensate drain age - Evaporator can be placed at a greater distance	- Increased noise level due to com pressor indoors possible - Oil return may be problematic under certain conditions - Heat loss through connecting line
Air	Air-to-water heat pump with split brine heat exchanger	- No risk of freezing - No heat loss through connecting pipes - Outdoor unit can be placed far ther from the house (good for noise)	- Additional circulation pump required - Losses due to additional heat ex changer - Foundation and frost-free condensate drain required
Air	Exhaust air-to-water heat pump	- Basement is dehumidified - Small space requirement - Relatively low acquisition costs	- Possibly high noise level in the building
Air	Air-to-air heat pump	- No water-carrying components - Wall or floor installation of the HP possible	- Expertise required according to the F-Gas Regulation - Heat losses through the connecting line

INDEXES

BIBLIOGRAPHY

[1] T. /. Schenker, Wärmepumpen / Wärmepumpenanlagen, Vogel Buchverlag, 2. Auflage 2014.

[2] Seifert, Effizienter Betrieb von Wärmepumpen-Anlagen, VDE Verlag, 2. Auflage 2020.

[3] Seifert, Wärmepumpen für Heizung und Warmwasser, 2022.

[4] BWP, Ratgeber Effizienzhaus mit Wärmepumpe, 2022.

[5] BWP, Das EU-Energielabel in der Praxis, 2015.

[6] BAFA, Der Energieausweis, 2020.

[7] DIN V 18599-1, Energetische Bewertung von Gebäuden - Berechnung des Nutz-, End- und Primärenergiebedarfs für Heizung, Kühlung, Lüftung, Trinkwarmwasser und Beleuchtung - Teil 1: Allgemeine Bilanzierungsverfahren, Begriffe, Zonierung und Bewertung der Energieträger, 2018.

[8] DIN EN 14511, Luftkonditionierer, Flüssigkeitskühlsätze und Wärmepumpen für die Raumbeheizung und -kühlung und Prozess-Kühler mit elektrisch angetriebenen Verdichtern - Teil 1: Begriffe, 2019.

[9] VDI 4650 Blatt 1, Berechnung der Jahresarbeitszahl von Wärmepumpenanalgen - Elektrowärmepumpen zur Raumheizung und Trinkwassererwärmung, 2019.

[10] DIN V 4701 -10 , Energetische Bewertung heiz- und raumlufttechnischer Anlagen - Teil 10: Heizung, Trinkwassererwärmung, Lüftung (zurückgezogen), 2003.

[11] „www.energieheld.ch," 2023. [Online].

[12] BWP, Leitfaden Trinkwassererwärmung, 2019.

[13] Gebäudeenergiegesetz, GEG, 2020.

[14] Viessmann, Planungshandbuch Wärmepumpen, 2011.

[15] Vaillant, Planungsinformation, 2019.

[16] S. Eltron, Planung und Installation, 2021.

[17] BWP, Kundenratgeber Erdwärme, 2021.

[18] VDI 4640 Blatt 1, Thermische Nutzung des Untergrunds - Grundlagen, Genehmigungen,

Umweltaspekte, 2010.

[19] Bundesamt Geothermie , Tipps für Hausbesitzer und Bauherrren, 2019.

[20] Energie Schweiz, Die Wärme-Pumpen-Heizung, 2018.

[21] „www.geothermie.de," 2023. [Online].

[22] Glen Dimplex Thermal Solutions, System Zero, 2017.

[23] NTB, Dr. Cordin Arpagaus, Hochtemperatur-Wärmepumpen für industrielle Anwendung, 2019.

[24] BWP, Ratgeber Energie aus Abwasser, 2019.

[25] HK-Gebäudetechnik, Eis für das Wärmen und Kühlen von Gebäuden, 2019.

[26] DIN/TS 12831-1, Verfahren zur Berechnung der Raumheizlast - Teil 1: Nationale Ergänzung zur DIN EN 12831-1, 2020.

[27] VDI 4645, Heizungsanlagen mit Wärmepumpen in Ein- und Mehrfamilienhäusern, 2018.

[28] DIN EN 12831 Beiblätter, Energetische Bewertung von Gebäuden, 2008-2020.

[29] DIN EN ISO 7730, Ergonomie der thermischen Umgebung - Analytische Bestimmung und Interpretation der thmerischen Behaglichkeit durch Berechnung des PMV- und des PPD-Indexes und Kriterien der lokalen thmerischen Behaglichkeit, 2006.

[30] VDI 2078, Berechnung der thermischen Lasten und Raumtemperaturen, 2015.

[31] Variotherm, Fussbodenheizung Trockenbau 20mm - Die superschlanke VarioKomp.

[32] „www.kesselheld.de," 2023. [Online].

[33] VDI 4640 Blatt 5, Thermische Nutzung des Untergrunds - Thermal-Rsponse-Test (TRT), 2020.

[34] VDI 4640 Blatt 2, Thermische Nutzung des Untergrunds - Erdgekoppelte Wärmepumpenanlagen, 2019.

[35] VDI 4640 Blatt 4, Thermische Nutzung des Untergrunds - Direkte Nutzungen, 2004.

[36] BWP, Informationsblatt Nr. 73 - Trinkwassererwärmung mit Trinkwasser-Wärmepumpe und Photovoltaik, 2020.

[37] EN 15450, Heizungsanlagen in Gebäuden - Planung von Heizungsanalgen mit Wärmepumpen, 2007.

[38] DIN 4708-2, Zentrale Wassererwärmungsanlagen; Regeln zur Ermittlung des Wärmebedarfs zur Erwärmung von Trinkwasser in Wohngebäuden, 1994.

[39] DIN EN 12831-3, Energetische Bewertung von Gebäuden - Verfahren zur Berechnung der Norm-Heizlast Teil 3: Trinkwassererwärmungsanlagen, Heizlast und Bedarfsbestimmung, 2017.

[40] DIN EN 12831-1, Energetische Bewertung von Gebäuden - Verfahren zur Berechnung der Norm-Heizlast - Teil 1: Raumheizlast, Modul M3-3, 2017.

[41] BWP, Leitfaden Hydraulik, 2019.

[42] BWP, Informationsblatt Nr. 62 - Inspektion, Wartung und Optimierung von Heizungsanlagen mit Wärmepumpe, 2015.

[43] BWP, Leitfaden Aussenaufstellung von Wärmepumpen mit brennbaren Kältemitteln, 2021.

[44] DIN EN 378, Kälteanlagen und Wärmepumpen, -.

[45] BWP, Praxisratgeber Modernisierung mit Wärmepumpe, 2021.

[46] „www.waermepumpe.de," [Online].

[47] BWP, Informationsblatt Nr. 68 - System Photovoltaik, Wärmepumpe und Speicher, 2017.

[48] BWP, Informationsblatt Nr. 57 - Bivalente Wärmepumpen-Systeme, 2014.

[49] BVF, Richtlinie 14 - Flächenheizungen und Flächenkühlungen im Niedrigstenergiegebäude, 2020.

[50] BWP, Informationsblatt Nr. 70 - Planung und Auslegung des Systems PV-Anlage, Wärmepumpe und Speicherung, 2018.

Figure	Figure Caption	Source of the figure
1	Eureka-Luft (EU-L) series by LAMBDA Wärmepumpen GmbH	Received photo and with permission from LAMBDA Wärmepumpen GmbH.
2	Types of Energy	Created in accordance with BWP, Guide Efficiency House with Heat Pump, 2022.
3	Illustration of COP and SPF (HSPF) throughout the year.	Created in reference to www.energie-experten.org.
4	Boundary limits on WPA and WPHA for calculating the (Heating)	Created in accordance with VDI 4650 Part 1, Calculation of the seasonal coefficient of performance of heat pumps Electric heat pumps for space heating and domestic hot water, 2019.
5	Display, heat pump, exemplified	Specially designed.
6	Principle of Operation: Heat Pump	Self-created based on BWP Guide for Modernization with Heat Pumps.
7	Structure of a Heat Pump	Created in accordance with Vaillant Planning Information, 2019.
8	Thermodynamic Cycle of a Compression Heat Pump	Specially designed.
9	Evaporator, exemplary	Specially designed.
10	Compressor, exemplary	Specially designed.
11	Expansion valve, exemplary	Specially designed.
12	Functional principle cooling and heating	Specially designed.
13	Availability-to-efficiency ratio	Created in accordance with Viessmann, Heat Pump Planning Handbook, 2011.
14	Structure of groundwater heat source	Created based on BWP, Customer Guide Geothermal Energy, 2021.
15	System Groundwater Heat Source	Created in accordance with Viessmann, Heat Pump Planning Handbook, 2011
16	Seasonal and borehole depth-dependent temperature variations of the ground	Custom-made with reference to Vaillant, Planning Information, 2019.
17	Ground Source Heat Pump System	Custom-made with reference to BWP, Customer Guide to Geothermal Energy, 2021.
18	Different ground loops	Custom-made with reference to Bundesamt Geothermie, Tips for Homeowners and Builders, 2019.
19	Borehole, exemplary	Specially designed.
20	Artesian well	Custom-made based on the respective depiction on Wikipedia.
21	Distribution box, located outside in a shaft	Specially designed.
22	Horizontally laid ring trench collectors	Specially designed.
23	Geothermal baskets	Custom-made with reference to VDI 4640 Part 2, Thermal Use of the Subsurface - Ground-coupled Heat Pump Systems, 2019.
24	Example of a trench collector	Custom-made with reference to VDI 4640 Part 2, Thermal Use of the Subsurface - Ground-coupled Heat Pump Systems, 2019.
25	Example of an energy fence	Specially designed.
26	Monoblock-design, Eureka-Luft (EU-L) - series by LAMBDA Wärmepumpen GmbH	Custom-made with reference to www.vaillant.ch.
27	Operating principle of monoblock design, outdoor installation	Custom-made with reference to www.vaillant.ch.

Figure	Figure Caption	Source of the figure
28	Operating principle of monoblock design, indoor installation	Specially designed.
29	Operating principle of split design	Custom-made with reference to www.vaillant.ch.
30	Operating principle of split construction	Custom-made with reference to Vaillant, Planning Information, 2019.
31	Air circulation/hot water heat pump (left) and exhaust air/hot water heat pump (right)	Created in accordance with Viessmann, Heat Pump Planning Handbook, 2011
32	Functioning principle of a heat pump with multiple heat sources	Specially designed.
33	Classification of heat pumps according to temperature	Custom-made with reference to NTB, Dr. Cordin Arpagaus, High-Temperature Heat Pumps for Industrial Applications, 2019.
34	Illustration of district heating and cold district heating	Custom-made with reference to BWP, Guide to Energy from Wastewater, 2019.
35	Illustration of the Concept of an Ice Storage System	Custom-made with reference to HK-Gebäudetechnik, Ice for Heating and Cooling of Buildings, 2019.
36	Representation of Monovalent Operating Mode	Custom-made with reference to Vaillant, Planning Information, 2019.
37	Representation of Bivalent Parallel Operating Mode; TA = Activation Temperature for Supplementary Heating	Custom-made with reference to Vaillant, Planning Information, 2019.
38	Representation of Bivalent Alternative Operating Mode; TA = Activation Temperature for Second Heat Generator and Deactivation of First Heat Generator	Custom-made with reference to Vaillant, Planning Information, 2019.
39	Representation of Bivalent Partial Parallel Operating Mode	Custom-made with reference to Vaillant, Planning Information, 2019.
40	Series Buffer Tank in the Flow	Created in accordance with Viessmann, Heat Pump Planning Handbook, 2011
41	Series Buffer Tank in the Return	Created in accordance with Viessmann, Heat Pump Planning Handbook, 2011
42	Parallel Buffer Tank	Created in accordance with Viessmann, Heat Pump Planning Handbook, 2011
43	Open Hot Water Tank (Top) and Closed Hot Water Tank (Bottom) Principle	Custom-made with reference to www.aeg-haustechnik.de.
44	Monovalent Domestic Hot Water Preparation System	Created in accordance with Viessmann, Heat Pump Planning Handbook, 2011
45	Monoenergetic Domestic Hot Water Preparation System	Created in accordance with Viessmann, Heat Pump Planning Handbook, 2011
46	Bivalent Domestic Hot Water Preparation System	Created in accordance with Viessmann, Heat Pump Planning Handbook, 2011
47	Storage Charging System	Created in accordance with Viessmann, Heat Pump Planning Handbook, 2011
48	Combination tank with internal heat exchanger	Created in accordance with Viessmann, Heat Pump Planning Handbook, 2011
49	Combination tank with tank-in-tank system	Created in accordance with Viessmann, Heat Pump Planning Handbook, 2011
50	Step-By-Step Guide of a heat pump system, overall view	"Specifically designed in reference to VDI 4645."

Figure	Figure Caption	Source of the figure
51	Overview of responsibilities in heat pump systems	"Specifically designed referencing Seifert, Efficient Operation of Heat Pump Systems, VDE Publishing, 2nd Edition 2020."
52	Planning scheme of a heat pump system, heating load calculation	"Specifically designed in reference to VDI 4645."
53	Planning scheme of a heat pump system, cooling mode	"Specifically designed in reference to VDI 4645."
54	Planning scheme of a heat pump system, design of heat transfer	"Specifically designed in reference to VDI 4645."
55	Underfloor heating in dry construction	Specially designed.
56	Underfloor heating in a wet system	Specially designed.
57	Principle of concrete core activation	Custom-made with reference to www.hausjournal.net.
58	Wall heating "ACULAR-M" by THERMOLUTZ GmbH & Co. Heizungstechnik KG	Document Ceiling Heating and Cooling Modules - SYSTEM "ACULAR-M" with permission from THERMOLUTZ GmbH & Co. Heating Heizungstechnik KG.
59	Ceiling heating and cooling modules "SYSTEM acular DECKE" by THERMOLUTZ GmbH & Co. Heizungstechnik KG	Document Ceiling Heating and Cooling Modules - SYSTEM acular DECKE with permission from THERMOLUTZ GmbH & Co. Heating Heizungstechnik KG.
60	Carpet pictogram: Suitable for underfloor heating floors	Specially designed.
61	Step-By-Step Guide of a heat pump system, selection of the heat source	Specially designed based on VDI 4645.
62	Step-By-Step Guide of a heat pump system, determination of hot water heating energy demand	Specially designed based on VDI 4645.
63	Step-By-Step Guide of a heat pump system, heating concept	Specially designed based on VDI 4645.
64	Step-By-Step Guide of a heat pump system, hydraulic concept	Specially designed based on VDI 4645.
65	Step-By-Step Guide of a heat pump system, sizing and selection of the heat pump	Specially designed based on VDI 4645.
66	Heat pump cascade by WOLF GmbH	Photo obtained with permission from WOLF GmbH.
67	Step-By-Step Guide of a heat pump system, control system	Specially designed based on VDI 4645.
68	Step-By-Step Guide of a heat pump system, economic analysis	Specially designed based on VDI 4645.
69	Step-By-Step Guide of a heat pump system, establishment of system control	Specially designed based on VDI 4645.
70	Step-By-Step Guide of a heat pump system, selection of the proposal and awarding of the contract	Specially designed based on VDI 4645.
71	Exemplary circulation pump	Specially designed.
72	Exemplary insulation of the heating pipes	Specially designed.
73	Exemplary electricity storage	Specially designed.
74	Interaction of the three components: photovoltaic system, heat pump, and storage	Specially created in reference to BWP - Information Part No. 70 - Planning and Design of the PV System, Heat Pump, and Storage System.

Figure	Figure Caption	Source of the figure
75	Energy manager display: Early morning	Specially created in reference to BWP - Information Part No. 68 - Photovoltaic System, Heat Pump, and Storage
76	Energy manager display: Morning	Specially created in reference to BWP - Information Part No. 68 - Photovoltaic System, Heat Pump, and Storage
77	Energy manager display: Afternoon	Specially created in reference to BWP - Information Part No. 68 - Photovoltaic System, Heat Pump, and Storage
78	Energy manager display: Evening	Specially created in reference to BWP - Information Part No. 68 - Photovoltaic System, Heat Pump, and Storage
79	Exemplary app	Specially designed.
80	Eureka-Luft (EU-L) series by LAMBDA Wärmepumpen GmbH	Received photo and with permission from LAMBDA Wärmepumpen GmbH.
81	Step-By-Step Guide of a heat pump system, overall view, Appendix A	Specially designed based on VDI 4645.

LIST OF TABLES

IMPRINT

Sebastian Ziegler
Im Fuchswinkel 6B
5032 Aarau-Rohr, Switzerland
E-Mail: Published.Just.4u@gmail.com

INDEX

IMPRINT

Please see imprint on page 153.

Made in United States
North Haven, CT
04 November 2026

21770479R00092